In Search of William Gascoigne

Astrophysics and Space Science Library

For further volumes:
http://www.springer.com/series/5664

David Sellers

In Search of William Gascoigne

Seventeenth Century Astronomer

Springer

David Sellers
Leeds
UK

ISSN 0067–0057
ISBN 978-1-4614-4096-3 ISBN 978-1-4614-4097-0 (eBook)
DOI 10.1007/978-1-4614-4097-0
Springer New York Heidelberg Dordrecht London

Library of Congress Control Number: 2012940745

Printed on acid-free paper

Springer is part of Springer Science+Business Media (www.springer.com)

To Jane, Andrew, and Graham

Acknowledgements

This book could not have been written without access to many excellent public and academic libraries and their dedicated staff. In particular, the collections of Leeds University Library, Leeds City Library, the Bodleian Library (Oxford), the British Library, Cambridge University Library, Chetham's Library (Manchester), and the Royal Society Library have been indispensable.

Over the years that this book has been in preparation, I have received kind assistance from many individuals. My thanks are especially due to Wilbur Applebaum, who generously shared with me some of his transcriptions of Flamsteed's handwritten copies of the Gascoigne-Crabtree correspondence; Chris Webb of the Borthwick Institute in York, who expertly answered my questions about arcane aspects of seventeenth-century court records; Christine Brown and Philippa Hoskin, who assisted with the translation of some Latin texts; and Bill Sheehan, who made useful comments on the drafts of some of the early chapters and gave valuable encouragement.

For assistance with the provision of some key illustrations or images, I thank Vanessa Mitchell (Towneley Hall Museum and Art Gallery), Dr. Beatrix Schönewald (Stadtmuseum Ingolstadt), Jo Hopkins and Jennifer Kren (Royal Society), Michael Powell (Chetham's Library), Kerry Magruder (University of Oklahoma Libraries), Chris Sheppard (Leeds University Library), and Don Manning (Cambridge University Library).

I am grateful to the staff at Springer for seeing the book through to publication – especially Harry Blom for having confidence in the project and Jennifer Satten for patiently guiding me through the process.

Finally, I thank my wife Jane and my sons Andrew and Graham for tolerating my interest in (or obsession with!) Gascoigne's story for many years and for reading parts of the manuscript or acting as an instant 'sounding' board for much of the material which has found its way into the book (and lots that rightly hasn't).

Contents

Part I
The Discovery of William Gascoigne

Chapter 1
Introduction

1638 December 10th, Middleton near Leeds[1]: The unmade track known locally as Town Street is deserted. At an hour past midnight in the depth of winter most folk are asleep in bed. No lamps light the track. Only the light of the full moon, this night lacking its usual vigour, illuminates the path. A strong wind sends thick clouds scudding across the sky, repeatedly plunging the terrain into darkness.

Low in the valley of the River Aire to the north is the township of Leeds, scarcely visible against the backdrop of the hillside beyond.

To the west can be seen the imposing bulk of Middleton Hall, residence of Sir Ferdinando Leigh, Gentleman of the Privy Chamber to Charles I and Governor of the Isle of Man.[2]

Here at the eastern end of Town Street is New Hall: a less extravagant affair, though still clearly the abode of local gentry. The hall has two wings with square mullioned windows, behind which are low rooms beneath massive oak timbers.[3] Nine chimneys proclaim the affluence of the inhabitants.[4]

A lamp swings to and fro within a chamber overlooking the track and just inside the window, silhouetted against the pale light of the lamp, stands the figure of a young man. He appears deceptively bulky—the effect of wearing many layers of clothing against the chill night air—as he tries to adjust the orientation of a wedge-shaped instrument, at the same time peering along its edge at the southern sky. Intermittently, he casts a sharp glance towards a clock at the rear of the chamber and jots down a record of his observations.

This is William Gascoigne—26–year old eldest son of New Hall's owner, Henry Gascoigne, and nephew of Sir Ferdinando Leigh—and he is assiduously observing

---

[1] Dates are given, except where otherwise stated, according to the Julian Calendar that was used in England up to 1752 (see page 153).

[2] Taylor, R.V., *Biographia Leodiensis* (London, 1865), 90.

[3] Britten, B., et al., *Belle Isle*, Belle Isle Study Group (Leeds, 1985), 33.

[4] Hey, D., et al, *Yorkshire West Riding Hearth Tax Assessment Lady Day 1677* (London, 2007), 272, The New Hall, Middleton cum Thorpe.

D. Sellers, *In Search of William Gascoigne: Seventeenth Century Astronomer*, Astrophysics and Space Science Library 390, DOI 10.1007/978-1-4614-4097-0_1,

a total eclipse of the moon. By degrees, the shadow cast by the Earth completely envelops the moon's disk. After the eclipse is over he notes

> ...the rageing windes would not admit any observation by the sextant whereby I might have rectified our clock except one altitude of the moon taken out of a Chamber which was 56°06' at which time the clock was 1h.29' observed by a wier & the minute wheele purposely figured...[5]

It is extraordinary to find a sextant being used in this remote part of northern England, so far from the company of other experimenters, but William is a pioneer of such artefacts. Near to the hall is a barn full of machines by which his father hopes to achieve great things.[6] Perhaps William is 'a chip off the old block', since he too is fascinated by mechanical devices. He has become adept at devising astronomical instruments ... and is without peer in their use. Soon, he will invent the telescopic sight and the telescope micrometer. With them he will make astronomical measurements of unprecedented accuracy. His achievements are even more remarkable in view of his claim to spend in this pursuit only that time that his friends devote to their hawks and hounds.[7]

Tragedy, however, is waiting in the wings. Within 6 years he will be dead: Brutally slain on a battlefield not far from his home. His naked body will be thrown into a mass burial pit along with at least 4,000 other young men.

Nevertheless, by that time his astronomical measurements will have sufficed to ensure ultimate renown. Even 80 years later the opening pages of the monumental *Historiae Coelestis Brittanicae* of John Flamsteed, the first Astronomer Royal at Greenwich, will be dominated by William's measurements. Ultimately, his fame will spread to astronomers around the world and his place within the annals of astronomical history will be assured.

His name will be indelibly associated with those of two other young men with whom he will soon enjoy a brief collaboration: Jeremiah Horrocks of Liverpool and William Crabtree of Broughton, near Manchester. Unbeknown to Gascoigne, they too are presently observing this same eclipse, but from the other side of the Pennine hills. Together with him, they will blaze a trail for precision astronomy in England and for a brief period they will pioneer the most advanced astronomical techniques and ideas in Europe. The lives of all three will be lamentably short.

Horrocks will correctly predict the 1639 'transit' of the planet Venus across the face of the Sun, which he and Crabtree will successfully observe. As a result, their names will redound through the centuries. Transits of Venus occur in pairs, each one within the pair separated by 8 years from the other. Each pair is separated from the next, however, by more than a hundred years: time enough for several generations to live out their lives. In each new 'transit season' the immortal story of Horrocks

---

[5] The National Archives, RGO 1/40, f.19v. This is the 20–21 Dec 1638 (NS) eclipse.

[6] *Ibid.*, f.22r, Christopher Towneley told Flamsteed that Gascoigne's father 'was much given to mechanick by which hee hoped to performe no meane thinges & once promised to show all sorts of manifactures workeing by them'.

[7] Rigaud, S.P. and S.J., *Correspondence of scientific men of the seventeenth century* (Oxford, 1841), 34.

and Crabtree will be retold and their fame reinforced.[8] By contrast, the story of William Gascoigne, not having the benefit of a celestial reminder, will eventually fall into a singularly undeserved obscurity.

Middleton on the southern outskirts of Leeds, in Yorkshire, has been completely transformed since Gascoigne's day. In the 1630s and 1640s, during his youth and early adulthood, the population was small and few buildings of any substance made their mark on the area. The landscape was pocked with tiny pits dug down to the coal seams which lay just beneath the surface. The presence of coal was a factor in the wealth of the Gascoignes of Thorp–on–the–Hill, William's branch of the family: the wealth that afforded him the leisure to immerse himself in astronomy and to purchase the latest continental books on the subject.

By the late eighteenth century things had hardly changed. A string of modest dwellings had been erected along Town Street, but New Hall—devoid now of Gascoignes—still dominated the eastern end, where it stood at the brow of the hill whose northern flank swept sharply down towards a solitary windmill. In the 1850s, when the first detailed map of the area was published by the Ordnance Survey, it confirmed the essentially rural setting of New Hall. Numerous shallow coal workings were recorded on the map. The burgeoning townships of Leeds, Holbeck and Hunslet, nestling in the Aire valley 6 or 7 km to the north had begun to place demands for fuel supply that would soon transform Middleton. Already, as early as 1812, the extraction of Middleton coal had reached a scale that justified the building of a local colliery railway: the first commercial steam locomotion undertaking in the world.

The increasing requirement of the mills and factories of Leeds for steam power resulted in much of the surrounding estate, known locally as *Belle Isle*, being turned over to coal mining interests. The pit head buildings and winding gear of the Middleton Broom Colliery occupied substantial land to the immediate northwest of a forlorn looking New Hall. By degrees, the pastoral prospects had been replaced by a panorama of coal-blackened, slag-strewn landscape.

In an 1871 guide to walkers, WS Banks of Wakefield, warned that

> at the foot of the Middleton hill lies a wretched-looking hamlet absurdly called "Belle Isle", if we are to judge it by its present condition; but if the name ever was correct that must have been conferred in happier days than ours, for now neither of the words by which it is known is in any sense appropriate. It is rickety and dirty, and almost buried beneath tramway embankments. The "New Hall" of Middleton stood until lately on the eastern part of the township, but a year or two ago it became dilapidated, and part of the roof fell in, and the house was taken down. A modern plain brick dwelling now stands on the site. Notwithstanding its name, *New* Hall, it was an old building, and remarkable as the residence of William Gascoigne.[9]

[8] Sellers, David, *The Transit of Venus: the quest to find the true distance of the Sun* (Leeds, 2001); Sheehan, William & Westfall, John, *The Transits of Venus* (New York, 2004); Maor, Eli, *Venus in Transit* (Princeton & Oxford, 2004); Aughton, Peter, *The Transit of Venus: the brief, brilliant life of Jeremiah Horrocks* (London, 2004); Lomb, Nick, *The Transit of Venus: 1631 to the present* (Sydney, 2011); Applebaum, Wilbur, *Venus seen on the Sun: the first observation of a transit of Venus by Jeremiah Horrocks* (Leiden, 2012).

[9] Banks, W.S., *Walks in Yorkshire: Wakefield and its Neighbourhood* (London & Wakefield, 1871), 146.

Fig. 1.1 The only local memorial to William Gascoigne (Author's collection)

The new building on the site became known as New Hall Farm.

By the 1920s one of the first municipal housing estates had been established in Belle Isle and the metalled roads, having long since replaced the unmade tracks of Gascoigne's day, now sported tramlines. As the sprawling housing estates of Belle Isle and Middleton extended during the twentieth century, eventually displacing the colliery, New Hall Farm was demolished and a simple brick building, accommodating several small local stores, now stands in its place.

Little vestige of Gascoigne's Middleton remains. Few residents have heard of the young astronomer who lived in these parts so long ago. The local high school that briefly bore his name has now closed and the only local memorial to his achievements is the dilapidated sign above the 'William Gascoigne Youth Club' (Fig. 1.1).

Unearthing and piecing together information about William Gascoigne is difficult, for few relevant contemporary documents exist. Scarcely any original copies of his correspondence have survived. Upon his death, most of his papers, books, instruments and other property were lost in the maelstrom of the civil war. Even the few letters that were rescued have subsequently disappeared. Most published biographies[10] have been necessarily brief—scarcely longer than a dictionary entry—and little of his work has ever been published. This book tells how Gascoigne's story was saved from total oblivion and, by exploring the archives, extends what we know.

[10] Perhaps the lengthiest account of William Gascoigne was provided in the 41-page pamphlet *Three North Country Astronomers*, by Allan Chapman (Farnworth, 1982). Other notable accounts include those by: Frances Willmoth in *Encyclopedia of the Scientific Revolution* (ed. Wilbur Applebaum) (New York & Abingdon, 2000), 255-6; Victor E.Thoren in *Dictionary of Scientific Biography* (ed Charles Coulston Gillespie) (New York, 1972), v.5, 278-9; Allan Chapman in *Oxford Dictionary of National Biography* (Oxford, 2004), v.21, 591–3, and in *Dividing the Circle* (Chichester, 1995), 35–45; R.V.Taylor in *Biographia Leodiensis* (London, 1865), 86–7.

Chapter 2
The Gascoignes of Thorp-on-the-Hill

William Gascoigne was the eldest son of Henry Gascoigne of Thorp–on–the–Hill near Middleton, in the parish of Rothwell. By the time of William's birth, Henry was living in comfortable circumstances, but at its outset, Henry's life had been scarred by tragedy. In 1589, aged only three, he became an orphan. In accordance with the dying wishes of their father, he and his sister, Elizabeth, were put into the care of their wealthy uncle, Richard Tempest of Tong.[1] In the Tempest household near Bradford brother and sister spent their early childhood years together, but by the time Henry was 12 they were separated as his custody and wardship was made over to his future father-in-law, William Cartwright of York.[2] An unsettling transfer—some would consider it heartless—but this was seventeenth century England: the wardship of such a minor was a commodity to be competed for. Sometimes it would be granted by the Court of Wards to a complete outsider, simply the highest bidder, keen to gain income and advantage from the orphan's assets. Not until 1611 would the law be changed to ensure that the next of kin had a prior right to the wardship.

As a ward, Henry brought benefits: Though bereft of parents, he was far from being destitute. His father, John, had bequeathed to Henry and Elizabeth 'manors, messuages, milnes, mines, land, tenements, rents [and] woods'.[3] These were assigned to William Cartwright 'until the children become of age or for 15 years'.[4]

---

[1] West Yorkshire Archive Service (Bradford), *Records of the Tempest Family of Tong Hall*, Tong/3/136, '.... *On the death of the said John Gascoigne, the said Richard Tempest becomes guardian to the two children [Henry and Elizabeth]* ....', 24 September 1598.

[2] *Ibid*., Tong/3/137, 29 September 1598.

[3] A *messuage* was a dwelling house, along with the land around it and any outbuildings. A *tenement* was property held by one person from another.

[4] *Ibid*., Tong/8/136, 24 September 1598. Ultimately, Henry—the heir to the estate—was to dispute the amount owed to him by Richard. This led to litigation over the sum of £1,980, which Henry claimed. The dispute was not fully settled until January 1610 (ref: Tong/7c/1). See also Cliffe,J.T., *The Yorkshire Gentry from the Reformation to the Civil War* (London, 1969), 377. The charges levied by the Court of Wards when it granted the wardship of Henry Gascoigne in 1591, whose estate was worth an annual income of £360, were: Fine, £4. 6s. 8d.; Rent, £3. 9s. 1d (Source: PRO, Court of Wards, Miscellaneous Books, Wards 9/cxviii/f.307 and cccxlviii (no pagination), quoted by Cliffe, J.T., *Op.Cit*., 133).

D. Sellers, *In Search of William Gascoigne: Seventeenth Century Astronomer*, Astrophysics and Space Science Library 390, DOI 10.1007/978-1-4614-4097-0_2,

Fig. 2.1 St. Michael le Belfrey church in York (Author's collection)

It was clear that Henry would be a 'gentleman'. He would not have to earn his living by labour: He would derive a very comfortable income from land, mines and other assets. In short, from the labour of others. He would also inherit his father's right to a coat of arms and would therefore be known as Henry Gascoigne, *Armiger* (sometimes rendered in English as 'Esquire').

At the tender age of 13 Henry married Cartwright's daughter, Jane.[5] The wedding took place at the parish church of St.Michael le Belfrey (Fig. 2.1), adjacent to York Minister, on 5th January 1600.[6] The church register has the simple entry: 'Henrye gascoyne [and] Jane Cartwright was maryed w'th a Licence, the vth of January'. A licence wasn't cheap—only the gentry and nobility could normally afford one—but by the use of this device the couple avoided the need for banns to be read in the parishes of the bride and groom on three successive Sundays prior to the wedding. Thus, it provided a means of marrying in a hurry or keeping wedding arrangements private. The reason for the couple resorting to this procedure is not recorded.

Such a juvenile marriage as that of Henry and Jane, at the onset of puberty, was not unusual amongst the gentry and aristocracy. The match would be a calculated business transaction, arranged by parents, rather than the outcome of a romance. Very young newly-weds would not be expected to cohabit. By the time William—their first son and the subject of this story—was born, Henry and Jane had been

[5] The Registers of St.Michael le Belfrey, York, Part I (1565-1653), transcribed by Francis Collins (Yorkshire Parish Register Society, 1899), 84, *'WEDDINGES in Anno D'ni 1599 .... Henrye gascoyne [and] Jane Cartwright was maryed w'th a Licence the vth of January'*. The use of a licence possibly shows that Henry and Jane were marrying outside their own parish. Henry's orphan status might also have been relevant.

[6] The narrative year is the one commencing on 1st January.

Fig. 2.2 The Gascoigne coat of arms (Author's collection)

married for more than a decade. Although accounts differ, it appears that he was preceded by two sisters and, in due course, was followed by one sister and two brothers.

William's mother died when he was only 5 years old. She was buried at Rothwell parish church on the last day of August 1617—scarcely 8 months after the birth of her youngest son.[7] Not until 12 years after the burial of his childhood bride did Henry take a new partner. In July 1629, at the same church, he married Grace Thomas, the daughter of Richard Thomas of Thorp–on–the–Hill. In due course this new relationship produced several more children.

The preservation of information about the lineage of William Gascoigne has much to do with the fact that his ancestors were allowed to bear heraldic coats of arms—those exuberant, often colourful, personal adornments that marked out the privileged ranks of feudal society. Heraldry originated in the twelfth century as the use of distinctively painted shields to identify knights in armour on the battlefield. Shields would be passed down within families from father to son.

From the time of Henry I the right to grant arms was reserved to the sovereign. The inheritance of arms was governed by the Laws of Arms and the granting of them was delegated to the 'Kings of Arms'. The work of designing them and recording the relevant lines of descent was the job of the 'heralds'. It was even their job to indentify the corpses of those slaughtered on the battlefield by checking the heraldic insignia on their armour. The coat of arms of the Gascoignes was rather simple—the severed head of a golden conger eel (Fig. 2.2).

Eventually, many coats of arms became more intricate or elaborate and all made their way from armour to more peaceful devices—such as signet rings and documents.

From 1530 to 1686 heralds embarked on numerous visits to the regions of England and Wales for the purpose of ensuring that anyone using a coat of arms was properly entitled to it through genealogical descent. Thus, the legitimate bearers of

[7] The Register of Rothwell Parish Church, …., '1617 Burials … Magistra Jana uxor Henrici Gascoigne armigeri sepulta fuit ultimo die Augusti.' [Teacher/Lady Jane wife of Henry Gascoigne Esquire was buried on the last day of August. Note that 'magistra' may be a transcription error. Wheater, *Gentleman's Magazine* (London, 1863), 761, gives the register wording as 'Margaretta Jana'].

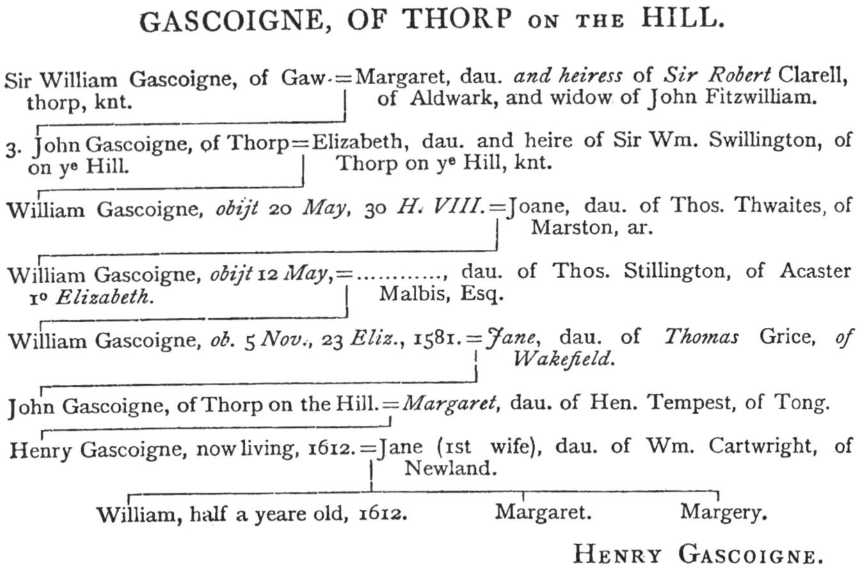

GASCOIGNE, OF THORP ON THE HILL.

Sir William Gascoigne, of Gawthorp, knt. = Margaret, dau. *and heiress* of *Sir Robert* Clarell, of Aldwark, and widow of John Fitzwilliam.

3. John Gascoigne, of Thorp on ye Hill. = Elizabeth, dau. and heire of Sir Wm. Swillington, of Thorp on ye Hill, knt.

William Gascoigne, *obijt* 20 *May*, 30 *H. VIII.* = Joane, dau. of Thos. Thwaites, of Marston, ar.

William Gascoigne, *obijt* 12 *May*, 1° *Elizabeth.* = ..., dau. of Thos. Stillington, of Acaster Malbis, Esq.

William Gascoigne, *ob.* 5 *Nov.*, 23 *Eliz.*, 1581. = *Jane*, dau. of *Thomas* Grice, *of Wakefield.*

John Gascoigne, of Thorp on the Hill. = *Margaret*, dau. of Hen. Tempest, of Tong.

Henry Gascoigne, now living, 1612. = Jane (1st wife), dau. of Wm. Cartwright, of Newland.

William, half a yeare old, 1612. Margaret. Margery.

HENRY GASCOIGNE.

Fig. 2.3 The family tree of the Gascoignes of Thorp-on-the-Hill (From *The Visitation of Yorkshire* (London, 1875), ed. J. Foster)

arms had their pedigrees, or family trees, recorded and scrutinised in the visitation records of the heralds.

Those considered to be unlawfully furnished with heraldry were subjected to public humiliation: forced to issue public disclaimers. Even the dead were not immune: any monument to the deceased, in church or graveyard, which bore suspect arms, would be torn down by the more zealous heralds.

When William Gascoigne was in his first year, Yorkshire received a 'Visitation' from one of the more prominent heralds: Sir Richard St. George. An active herald, a learned genealogist and the head of a dynasty of heralds, St. George had entered the College of Arms with the rank of Windsor in 1602 and became Norroy King of Arms in 1604.[8] In the course of his 1612 visitation, he meticulously recorded the pedigree of the Gascoignes of Thorp–on–the–Hill, from the half-year old William back through seven generations (see Fig. 2.3).

The pedigree of the Cartwrights of Newland, the family of William's mother, is also preserved in the same visitation records. Here, the young William is depicted at the end of a line of descent going through three generations.[9]

[8] Wagner, Sir Anthony, *Heralds of England: A History of the Office and College of Arms* (London, 1967), 226.

[9] Foster, Joseph, *The Visitation of Yorkshire made in the Years 1584/5, by Robert Glover, Somerset herald; to which is added The Subsequent Visitation made in 1612, by Richard St.George, Norroy King of Arms*, (London, 1875), 520.

The tracing of family trees and pedigrees was also a pre–occupation of various independent scholars, such as the respected Lofthouse antiquary, John Hopkinson, to whom a monument was erected in Rothwell parish church. Hopkinson had the rare distinction of being under the protection of both sides in the Civil War. Both Lord Fairfax and William, Marquesse of Newcastle—leaders of the Parliamentarians and Royalists respectively—issued special instructions to their troops not to molest Hopkinson or his family, so valuable did they consider his work.[10] His pedigree for the Gascoignes of Thorp–on–the–Hill was devised many decades later and continues the family tree beyond the death of William, the astronomer.

Despite occasional mistakes and contradictions, the pedigrees produced under the auspices of heraldic visits and by the efforts of antiquaries provide valuable information. Those arising from the 1612 *Visitation* were edited and privately printed in 1875 by Joseph Foster.

Since the mid-sixteenth century parish churches in England have been keeping records of baptisms, burials and marriages. A directive of Thomas Cromwell in 1538 required that such records be preserved in a 'secure coffer', or parish chest. It was not until the very end of the nineteenth century, however, that parish register societies started the work of transcribing the handwritten registers and publishing their content. These form perhaps a more reliable record than antiquaries' pedigrees.

Most accounts of William Gascoigne hitherto, have based themselves on secondary sources published before the 1870s, whose authors—not having easy access to either visitation records or parish registers—confined themselves to the pedigrees produced by antiquaries.

The various pedigrees are slightly at odds concerning William's siblings (see Fig. 2.4). The 1612 Visitation version only gives two sisters—Margaret and Margery—whereas Thoresby's 1715 version makes no mention of Margaret: Margery is given as Jane's only daughter, and two brothers make an appearance—Henry and John. The parish registers introduce further siblings.

Of the graves of the Gascoignes of Thorp-on-the-Hill no trace remains. The most durable should have been that of William's half-sister Ellinor, who died in 1663 and was buried under the floor of the chancel in Holy Trinity parish church in Rothwell, but the inscribed tombstone is no longer visible.[11]

The location and date of William's birth, despite freshly published registers, remains shrouded in obscurity. No one has so far found any Parish Register record

---

[10] Whitaker, Thomas Dunham, *Leodis and Elmete* (Leeds, 1816), 242.

[11] *Ibid.,* 242. The inscription read 'Hic jacet Hellenora Henrici Gascoigne Arm. Nuper de Thorp filia, uxor fidelis Arthuri Ingram jun. De Knottingley, trium liberorum charisima mater Anno xxiv mortem obiit, cujus pietatis & immaculatae virtutis exemplum non solum aevo praesenti sed praeter invidium futur ..... imitari debet 1669.' The Rothwell Parish Register, by contrast, records 'May 2, 1663 Elena uxr. Arthuri Ingram generosi de Thorpe sepult fuit die p'dic.'

	Children of **Jane** m.5 Jan 1600 b.31 Aug 1617						Children of **Grace** m.25 Jul 1629					
St George Visitation (H Gascoigne) – 1612	Margaret	Margery	William ½ yr old									
Ralph Thoresby – 1715			William d.Melton		Henry	John						Ellinor or Ellen
John Hopkinson – 1730	Margaret	Margery	William d.Melton		Henry d. London	John d. Oxford	Thomas		Richard		Margaret	Ellinor
Rothwell Parish Register (Christenings)				Elizabeth 1 Jan 1614	Henry 22 Sep 1615	John 1 Jan 1617	Thomas 13 Apr 1631	Maria 13 Sep 1632	Richard 26 Mar 1634	George 13 Jan 1636	Margaret 20 Sep 1637	Ellen 19 Nov 1639
Rothwell Parish Register (Burials)								Maria 12 Aug 1645	Richard 14 Feb 1641	George 3 Apr 1636	Margaret 19 Apr 1647	Ellen 2 May 1663

Fig. 2.4 The siblings of William Gascoigne according to various sources

of the christening or baptism of William, the putative astronomer. Likewise no unambiguous marriage record has been identified.[12]

We would have had no occasion, however, to look into any matter concerning the background of this young Yorkshire astronomer, had it not been for an intriguing discovery by a French scientist and his keenness to share that discovery with English friends.

[12] The Rothwell Parish Register has the following entry for 8th September 1630: "Willus Gascoigne et Rosomond Townend nupt' fuer'...". It is not certain that this refers to William Gascoigne of Thorp-on-the-Hill.

Chapter 3
The 'Discovery' of William Gascoigne

On 28th December 1666, more than two decades after the death of William Gascoigne, the French savant Adrien Auzout (1622–1691) picked up his pen to write to Henry Oldenburg (c.1619–1677), the publisher of the Royal Society's *Philosophical transactions*. He was eager to communicate to scientific friends across the Channel details of what he considered an entirely novel device that, together with Jean Picard, he had used to measure the angular diameter of the Sun, Moon and planets.

> We can take diameters to seconds, *he wrote*, because we can divide a foot into 29000 or 30000 parts, scarcely being able to mistake one single part, and we can be almost certain that we cannot be mistaken by more than 3 or 4 seconds. I will not tell you anything at the moment of my Observations, not having the time, but I can well assure you that the diameter of the sun was hardly smaller at its apogee than 31′ 37″ or 40″ and certainly was no less than 31′ 35″.[1]

The device being described was a telescope micrometer—an instrument which allowed an object in the telescope's field of view to be finely encompassed between moveable internal pointers, or hairs. The separation of the pointers could be read off some suitable scale and from this scale the separation could be converted to give an angle—the angle subtended at the eye by the observed object (Fig. 3.1).

The significance of being able to measure celestial angles with this accuracy would not have been lost on Oldenburg. The exact measurement of the changes in the apparent size of the Sun was a hot topic in astronomy. Provided sufficient precision could be achieved, one might be able to prove the shape of the Earth's orbit.

Auzout, born in Rouen in 1622, was a founding member of the Académie Royale des Sciences—France's scientific elite—and was one of Oldenburg's most reliable continental correspondents. There was every reason to trust the accuracy of his claims for the new instrument. Nevertheless, Auzout could not have anticipated the controversy that would greet his letter.

---

[1] Hall, A.Rupert and Marie Boas, *The Correspondence of Henry Oldenburg* (Madison, 1966), iii, 293, Letter from Adrien Auzout, 28 December 1666 (trans. DS).

D. Sellers, *In Search of William Gascoigne: Seventeenth Century Astronomer*, Astrophysics and Space Science Library 390, DOI 10.1007/978-1-4614-4097-0_3,

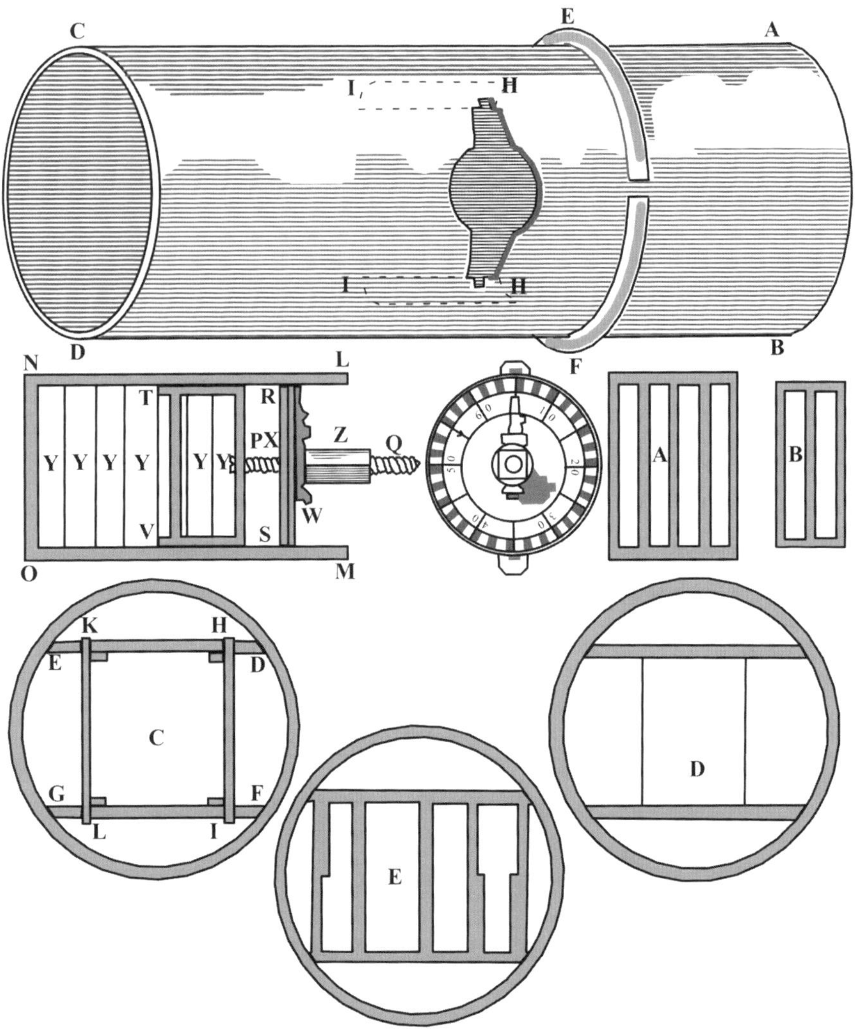

Fig. 3.1 Adrien Auzout's micrometer (Author's collection)

On 9th January 1667 Oldenburg read the letter to a gathering of the Royal Society. Robert Hooke and Christopher Wren responded vociferously—claiming that they had already known of similar instruments. They were asked to set down the details without delay 'that so it might be signified to the Parisian philosophers, that it was a thing not at all new among the English.'[2] Their claims seemed credible: The 1667 edition of Sprat's *History of the Royal Society* says of Wren that 'He has added many

[2] Birch, Thomas, *History of the Royal Society of London* (London, 1756), ii, 139.

sorts of Retes, Screws, and other devices to Telescopes, for taking small distances and apparent diameters to Seconds'.[3]

Possibly, the micrometer was truly a device whose time had come, since before long Oldenburg received a letter from another Frenchman, Petit, claiming the independent invention of an instrument for measuring the angular diameter of planets.[4] Christiaan Huygens and the Marquis Cornelis Malvasia had also, between 1659 and 1662, made angular measurements with devices based on similar principles.[5]

Oldenburg promptly published Auzout's letter in the *Philosophical transactions*.[6] When Richard Towneley (1629–1707) of Towneley, Lancashire, saw this he was dismayed. Not only was he aware of William Gascoigne's priority in relation to the invention of the ground-breaking micrometer, he was actually in possession of three such instruments made by Gascoigne a full quarter of a century earlier.

The Towneleys were prominent north-country Catholics: adherents of a faith which had been oppressed by the state for decades. Consequently, Richard's wide-ranging scientific interests had to be tempered by a low profile. But he did know people who moved in higher circles. One of them had been prescribing medication—purgative pills and the like—to the Towneley family for years. That was his good friend Henry Power (1626–1668), a Halifax physician. Their relationship went beyond patient-doctor matters. He and Power shared a love of science and together had carried out a series of pioneering measurements concerning atmospheric pressure. At Beacon Hill in Halifax and Pendle Hill in Lancashire they had shown how the height of a column of mercury changed as one ascended the hill. At Towneley Hall they had used a 32-ft column of water, in a tin tube attached to one of the turrets, to measure weight of the atmosphere. Power, recently elected to the newly formed Royal Society,[7] had set down an account of their observations in his book *Experimental Philosophy* (1664).[8]

---

[3] Sprat, Thomas, *History of the Royal Society* (London, 1667), 250. It is not clear when Sprat recorded the item about Wren measuring angular diameters, but Henry Oldenburg wrote on 24 November 1664 (to Robert Boyle) that 'Mr. Sprat intends to begin next week to print the History of our Institution'.

[4] Hall, A.Rupert and Marie Boas, *op.cit.*, iii, 418, Letter from Mons. Petit, 28 May 1667. Petit's letter of 12 March 1667, published in *Journal des sçavans* (16 May 1667, pp.102–108), acknowledged the work of Huygens, Malvasia, Auzout and Picard. In the next issue of the same journal (28 June 1667) Auzout gave an explanation of the differences between his 'machine' and that of Petit, along with a number of his observations.

[5] For excellent detailed accounts of the development of the micrometer, see: McKeon, Robert M., *Les débuts de l'astronomie de précision*, Physis, v.13 (1971), 225–288, and v.14 (1972), 221–242; Brooks, Randall C., *The development of the micrometer in the seventeenth, eighteenth and nineteenth centuries*, Journal for the History of Astronomy, v.22 (1991), 127–173.

[6] *Philosophical transactions*, i (1666), 373–75.

[7] Hunter, Michael, *The Royal Society and its Fellows, 1660–1700*, BSHA Monograph no.4, (Oxford, 1994), 154, 162. Power was elected on 26 February 1662 and re-elected on 1 July 1663.

[8] For the relationship between Towneley and Power, see also: Webster, C., *Richard Towneley (1629–1707), the Towneley group and seventeenth century science*, Transactions of the Historic Society of Lancashire and Cheshire, v.118 (1966), 51–76, esp. 66–68; Hughes, J.T., *Henry Power of Halifax* (Oxford, 2010), 57–63.

It was natural then that Towneley turned to his physician friend to express his concerns about Auzout's claim. As it happened, the doctor was already an admirer of Gascoigne. In his book he had described the young astronomer as 'our famous, and never to be forgotten Country-man, Master Gascoign of Midleton near Leeds, who was unfortunately slain in the Royal Service for His late Majesty; a Person he was of those strong Parts and Hopes, that not onely we, but the whole World of Learning suffered in the loss of him'.[9] His response to Towneley was forthright: The existence of the instruments should be brought without delay to the attention of the Royal Society to establish Gascoigne's priority. He persuaded Towneley to approach the Royal Society via a young fellow physician, Dr. William Croone (1633–1684). On 25th March 1667 Towneley wrote to Croone,

> Your former Civilities pinned to the Importunities of some of my friends, being particularly incouraged thereunto by Dr.Power, embolden me to give you this trouble, since I have not the good fortune to be known to the Publisher of the Philosophical Transactions. Finding in one of the last how much Monsr. Auzout esteems his Invention of dividing a foot into near 30000 parts, and taking thereby Angles to a very great Exactnesse; I am told I shall be look'd upon as a great wronger of our Nation, should I not let the World know, that I have out of some scatter'd Papers and Letters that formerly came to my hands, of a Gentleman of these parts, one Mr. Gascoigne, found out, that before the late unfortunate wars, he had not only devised an Instrument of as great a power as Monsr. Auzout's. but had also for some years made use of it, not onely for taking the Diameters of the Planets and distances upon Land, but had further endeavoured out of its' precisenesse, to gather many certainties in the Heavens: amongst which I shall onely mention one, the finding the Moon's distance from two observations, of her Horizontal and Meridional Diameters, which I the rather mention, because the French Astronomer in his Letter to the Publisher, esteems himself the first that took any such notice, as thereby to settle the Moon's Parallax; for our Countryman fully considered it before, and imparted it to an Acquaintance of his, who thereupon proposed to him, the difficulties that would arise in the Calculation, with considerations upon the strange niceties necessary to give him a certainty of what he desired. The very Instrument he first made, I have now by me, and two others more perfected by him, which doubtlesse he would have infinitely mended, had he not been slaine unfortunately in his late Majesty's service. He had a Treatise of Opticks ready for the Presse, but though I have used my utmost endeavour to retreive it, yet I have in that point been totally unsuccessfull; but some loose papers and letters I have, particularly about this Instrument, for taking of Angles, which was far from perfect. Neverthelesse I found it, so much to exceed all others, that I have used my Endeavour to make it exact and easily tractable, which above a year since I effected to my own desire; by the help of an ingenious and exact Watch-maker in these parts[10]: Since which time, I have not totally neglected it, but imployed it particularly in taking the distances, (as occasion served) of the Circum–jovialists,[11] towards a perfect settling their Motion. I shall onely say of it, that it is small, not exceeding in weight, not much in bignesse, an ordinary Pocket–watch, exactly marking about 40000. divisions in a foot, by the help of two Indexes, the one showing hundreds of Divisions, the other, Divisions of the hundred, every last division, in my small one containing 1/10 of an inch, and that so precisely, that as I use it,

[9] Power, Henry, *Experimental Philosophy* (London, 1664), 80.

[10] Towneley later said that the screws were made by the Lancashire-based father of the London clockmaker, Humphrey Adamson (fl. 1668–1682).

[11] The Galilean moons of Jupiter.

Fig. 3.2 Henry Oldenburg (From *The Record of the Royal Society of London*, Royal Society, 1940)

> there goes about 2½ divisions to a second, yet I have taken Land-angles several times to one Division, though for the reason Monsr. Auzout mentions, it be very hard to come to come to that exactnesse in the Heavens, viz. the swift motion of the Planets, yet to remedy that fault, I have devised a Rest, in which I find no small advantage, and not a little pleasing those persons who have seen it, being so easy to be made, and by the Observer managed, without the help of another; which second convenience my yet namelesse Instrument hath in great perfection, and is, by reason of its smallnesse and shape, easily applyable to any Telescope. Sr. if you think this Invention thus improved, worthy to be taken notice of by the curious, you may command a more perfect Description of it, or any of the Observations either Mr. Gascoigne or my self have made with it. A word or two in Answer to this, would very much oblige...[12]

The letter reached the Royal Society in July. By that time, however, the Secretary, Oldenburg (Fig. 3.2), was not in a position to do much about it, for he was incarcerated in the Tower of London, accused of 'dangerous desseins and practices', on account of his extensive correspondence with foreigners at a time when England was at war with the Dutch. The arrest warrant was signed on 20 June and the unlucky suspect was soon under lock and key, where he would remain until 26 August. Samuel Pepys speculated that suspicion had fallen on Oldenburg particularly because of his communication of news to 'a virtuoso in France, with whom he constantly corresponds in philosophical matters'[13]—thought by some to be none other than Auzout.

---

[12] Towneley, Richard, *Letter to Dr Croon, 25 March 1667*, Royal Society, EL/T/19.

[13] Pepys, S., *Diary and Correspondence of Samuel Pepys* (London, 1854, 4th edition), v.3, 170 (25 June 1667).

After Oldenburg's release, Towneley's letter, excluding the first sentence that mentioned the role of Power, eventually appeared in the *Philosophical transactions*.[14] About Gascoigne it revealed little: only that he was 'a Gentleman of these parts', that he had been 'slaine unfortunately in his late Majesty's service', and that he 'had a Treatise of Opticks ready for the Presse'. Beyond these discrete facts nothing was published about the identity of this intriguing astronomer.

Upon reading Towneley's letter claiming priority for Gascoigne in relation to the micrometer, Auzout was more intrigued than offended. Henri Justel (1620–1693)—one of Oldenburg's other French correspondents—reported that 'Mr Auzout behaved very well' in face of the controversy.[15] The French astronomer was certainly sceptical: Not only would the swift movement of heavenly bodies severely constrain the achievable accuracy, but the quality of the optics and the screw thread manufacture would also set a strict limit on the exactness of measurements. Auzout was, perhaps justifiably, surprised at Towneley's claim that his own version of the Gascoigne micrometer was graduated at 'above 2½ divisions to a second, yet I have taken Land-angles several times to one Division'. Auzout's reaction was that, if this was to be understood as he imagined, 'Mr Towneley must have seen a hair at nearly 120 ft, which is impossible'. Nevertheless, he looked forward to seeing a detailed description of the device. In the meantime, Justel assured Oldenburg,

> Mr Auzout … could not imagine your fellows wishing to print something that was untrue, that it might be that another person had drawn the same conclusions as well as himself, and that [his colleagues] should wait … because there was reason to hope for the description of the device and the observations of the moon made both by the person who is dead [i.e. Gascoigne] and Mr.Towneley.[16]

Should this 'yet namelesse instrument' prove of interest to the Royal Society, Towneley had added at the end of his letter, 'you may command a more perfect Description of it, or any of the Observations either Mr Gascoigne or my self have made with it'. He was better than his word: By 11th July 1667, Robert Hooke could report to the Society that Dr. Croone 'had received from Richard Towneley, esq., Mr. Gascoiyne's instrument for measuring the diameter of the stars[17] with great exactness'.[18] The instrument was shown to the Society a fortnight later, along with an instrument devised by Hooke for the same purpose. According to Thomas Birch's History of the Royal Society of London, Hooke's instrument was 'of more plain and easy use'.

The instrument sent by Towneley was remarkably compact. The mechanism was contained in a small brass box probably less than 100 mm in length. On the outside of the box, mounted at one end was a circular brass plate, whose outer edge carried a graduated scale with a 100 divisions. Through the centre of the plate and

[14] *Philosophical transactions*, ii (1667), 457–58.

[15] Hall, A.Rupert and Marie Boas, *op.cit.*, iii, 438, Letter from Justel, mid-June 1667.

[16] *Ibid.*, 438.

[17] The word *stars* is used here, in line with the practice of the times, to mean the planets.

[18] Birch, Thomas, o*p.cit.*, ii, 187–189.

passing through the length of the box was a carefully made single composite screw. The third of the screw's length nearest the graduated plate was somewhat thicker than the other two-thirds (to which it was joined, end-on) and had a finer thread (exactly half the pitch) (Fig. 3.3).

When the handle at the end of the screw was turned, the screw revolved, making one of the pointers (driven by the coarser screw) move along the box, away from the other pointer. As the pointer moved, it carried along a ruled bar that indicated the pointer's displacement in terms of the number of whole pitch-lengths. The part pitch-length of displacement (measured in hundredths) beyond the whole number was shown by a rotating index against the circular plate.

The third of the screw with the finer thread was attached by a threaded block to the removable cover of the micrometer. Since this cover carried the telescope mounting, the movement of this block caused the whole micrometer assembly to move half the distance of the movable pointer, but in the opposite direction. Thus, cleverly, the pair of pointers maintained a position in the centre of the field of view—both equidistant from the optical axis of the telescope.

Towneley claimed to have made Gascoigne's device more 'exact and easily tractable', but there remains no further description to reveal the nature of his modifications. Was it simply a question of refining the manufacture of the screw, or was it something more fundamental? This remains a mystery.[19]

Members returned several times to a discussion of the subject and resolved to let Towneley know that they judged Gascoigne's micrometer to be 'a very ingenious and useful contrivance'. Indeed, they would commission another to be made at their own expense. In due course a detailed description of it by Hooke (see Chap. 17) was read before the members on 4th November 1667[20] and printed in the Transactions in the same month—though not with the promised account of observations. No further information about the ingenious inventor was presented. Gascoigne remained in the shadows.

---

[19] Flamsteed, in a letter addressed to Towneley (25 January 1673) made reference to "your old micrometer in which *both the pointers* moved, had 1,307 parts to an inch, the first new screws 3,465, your late 3,415", but Hooke, in his published description of Gascoigne's micrometer (meaning the one given by Towneley), referred explicity to 'a fixed sight' (i.e. only one movable pointer). In his *Prolegomena*, or preface, to the *Historiae Coelestis*, Flamsteed said 'William Gascoigne noticed that all well-made screws should be of similar pitch and equal thickness; and therefore he made a pair of screws for the sides of his tube, to be inserted in such a way that their ends, by means of some indicator of revolutions, could approach or recede easily and equally in relation to each other' (Chapman, A. (ed.): Preface to John Flamsteed's Historia Coelestis Britannica (London 1982), 104). This is quite ambiguous, but still could mean counter-threaded screws joined along a common axis. Some writers have referred to Gascoigne's micrometer having two adjacent (parallel) screws, but this might be simply a misunderstanding of Hooke's description of two screws that are joined end-to-end. There does not appear to be any documentary evidence for such parallel screws.

[20] Hooke, Robert, *A Description Of an Instrument of dividing a Foot into many thousand parts and thereby measuring the Diameters of planets with great exactnesse etc*, 1667, Royal Society register book, RBO/3/65.

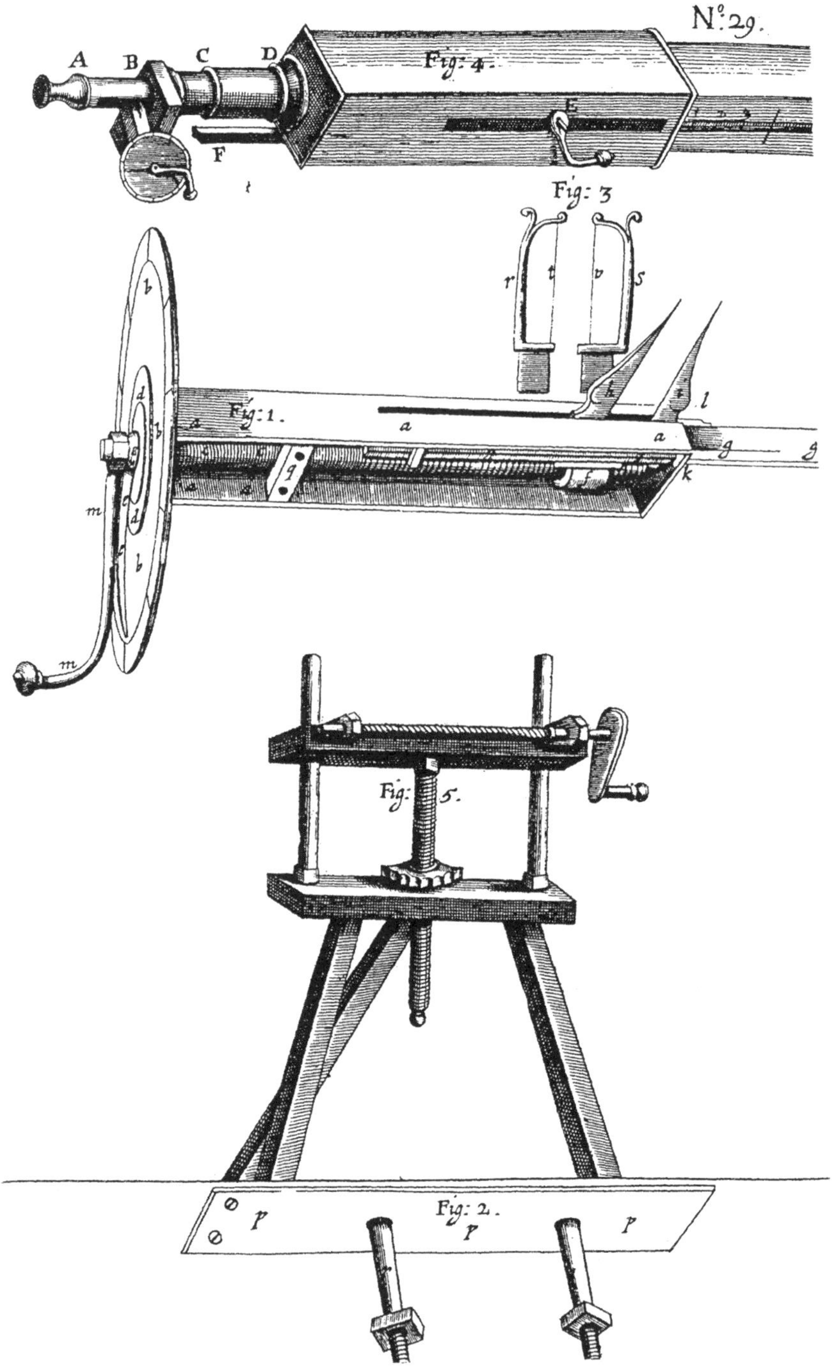

Fig. 3.3 Robert Hooke's drawing of Towneley's micrometer (From *Philosophical transactions*, ii (23–32), 541)

Chapter 4
A 'Light of the First Magnitude'

Not until 1675 did a few more meagre scraps about William Gascoigne appear in print. These came from Edward Sherburne (1616–1702), a translator, poet and hymn writer, who had lost a fortune supporting the King's cause in the Civil War. In the mid 1650s, whilst exiled in Paris, Sherburne began a labour of love that married his interest in astronomy and his skills as a poet. It was to take almost 20 years to complete the task: the poetic translation of the five volume, first-century poem *Astronomica*, written by the Roman poet Marcus Manilius. The finished result appeared as *The Sphere of Marcus Manilius made an English Poem* (Fig. 4.1).

Manilius had ventured, in the words of Sherburne, to inculcate

> Knowledge with Delight .... intending to exhibit to the Age wherein he lived the Rudiments of Astronomy, [choosing] to represent the same in a Poetical Dress, that so his Readers might be allured to relish with the greater Gusto the initiating Principles of a Science not easily acquired; and he thereby gain to himself the Repute which good Poets chiefly affect, of being able at once both to instruct and please.[1]

Sherburne had set the same ambitious aim for himself, but—going beyond the bounds of the poetic form—he also included an extensive Appendix with '*A Catalogue of Astronomers, Ancient and Modern*' in which he speaks of

> the most incomparable and ingenious Gentleman, William Gascoygne of Middleton in the County of York Esquire, who for some years before [making the acquaintance of Horrocks and Crabtree], had taken much pains in Astronomical Observations, and invented wayes to grind Glasses. He was the first that used two convex Glasses; had at that time a Tube that drew out fifteen foot in length, of his own working. The Object-glass was a Meniscus. He made several good Observations to be found in his Letters now in the Hands of Richard Townley of Townley in Lancashire Esquire, or in their extracts by Mr Jo. Flamsteed; and was questionless the first that ever exactly observed the Moon's Diameter in a Telescope by the help of Skrews, which were not before thought of. He invented wayes to measure small Angles by the Prospect-Glass to a second; and had he lived, he had certainly brought

[1] Sherburne, Edward, *The Sphere of Marcus Manilius made an English poem: with annotations and an astronomical appendix* (London, 1675), preface.

D. Sellers, *In Search of William Gascoigne: Seventeenth Century Astronomer*, Astrophysics and Space Science Library 390, DOI 10.1007/978-1-4614-4097-0_4,

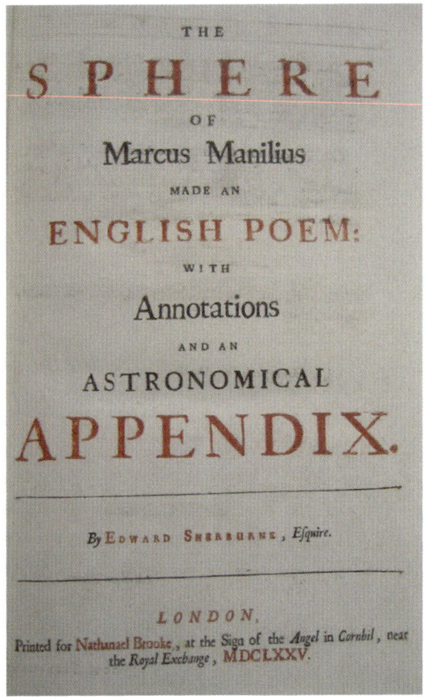
THE
SPHERE
OF
Marcus Manilius
MADE AN
ENGLISH POEM:
WITH
Annotations
AND AN
ASTRONOMICAL
APPENDIX.

By Edward Sherburne, Esquire.

LONDON,
Printed for Nathanael Brooke, at the Sign of the *Angel* in *Cornhil*, near the *Royal Exchange*, MDCLXXV.

Fig. 4.1 Sherburne's translation of The Sphere (Reproduced with the permission of Leeds University Library)

great perfection to Telescopes. Mr Crabtree taking a Journey into Yorkshire in the year 1639 writ thence to Mr Horrox, as followeth. The first thing Mr Gascoygne shewed me was a large Telescope amplified and adorned with new Inventions of his own, whereby he can take the Diameters of the Sun or Moon, or any small Angles in the Heavens, or upon the Earth most exactly through the Glass to a second. A device much desired, but little expected, etc. In the year 1641 Mr Gascoygne writ to Mr Crabtree, that he questioned not to provide an Instrument to take at once the Distances of three Points in the Periphery of the Earth's shade of the Moon, the Diameters of the Planets, etc. But he and all his excellent Inventions and Labours were lost by his Death, which was in his late Majesty's Service, in the Fight at Marston Moore.[2]

Gascoigne, Horrocks, Crabtree and William Milbourne, the mathematician, Sherburne added, were

lights of the first Magnitude, in the Northern Hemisphere, who were happily brought to the Acquaintance of one another by the means of Christopher Townley of Carr in Lancashire Esquire, who stuck not for any cost or labour to promote as well Astronomical as other Mathematical Studies by diligent correspondence kept and maintained with the learned Professors in those Sciences; upon which Account he was very dear to All the Four.[3]

[2] *Ibid*, 92, 105.

[3] *Ibid*, 92–93.

Known to some as 'the indefatigable transcriber', Christopher (1604–1674) was the uncle of Richard Towneley and had not only put Gascoigne, Crabtree and Horrocks in touch with each other, but, after their deaths, had also taken steps to collect and preserve many of their papers (Fig. 4.2).[4]

If Sherburne's account was to be trusted, a few more discrete facts thereby entered on the record for the first time: Gascoigne was an acquaintance of Crabtree, Horrocks, Milbourne and Towneley; made his own lenses and telescopes; and died at Marston Moor. We learn too that John Flamsteed (1646–1719) had transcribed extracts of Gascoigne's letters and observations.

Like the luminaries described by Sherburne, Flamsteed was also from the north of England. Born in 1646 at Denby in Derbyshire—the sickly son of a brewer. By his early twenties his skill in astronomical predictions had already come to the notice of the Royal Society and during a visit to London he had been introduced to Sir Jonas Moore, Surveyor-General of the Ordnance. Moore was to become a valuable patron. It was he who got the King to issue a warrant to Cambridge University on 14th May 1674 'to grant an MA degree to John Flamsteed, late of Jesus College, who has spent many years in the study of the liberal arts and sciences, and especially of astronomy, in which he has already made such useful observations as are well esteemed by persons eminently learned in that science'.[5] It was Moore too who later proposed that the young Flamsteed should be appointed as the first Astronomer Royal, in charge of the new Royal Observatory at Greenwich.

For some time before that prestigious appointment Flamsteed had been aware of Gascoigne's micrometer. Moreover, as a promising 24 year old in the summer of 1670, he had been presented with Towneley's variant of it by Jonas Moore. He resolved to find out more regarding the origins and use of the instrument at the earliest opportunity and relates in his autobiography that

> some affairs of my father's requiring it, in the month of June this year [1671] I made a journey into Lancashire, and called at Townly, to visit Mr. Christopher Townley,[6] who happened to be then in London. But, one of his domestics kindly received me, and showed me his instruments, and how his micrometer was fitted to his tubes: and from this time forward we often conferred by letters. I procured Mr. Gascoigne's and Crabtree's papers from him ….[7]
>
> In the spring of the year 1672 I excerpted several observations from Mr. Gascoigne's and Crabtree's letters, that had not yet been made public; which I had turned into Latin, and resolved to publish in the first volume of Celestial Observations taken at the Observatory.

---

[4] An account of the lives of William Crabtree and Jeremiah Horrocks is beyond the scope of this book. For more information, see: Chapman, Allan, *William Crabtree 1610–1644: Manchester's first mathematician* (Manchester, 1995); Radcliffe, Albert, *Brief Dawn: William Crabtree of Broughton and the Transit of Venus 24th November 1639* (Manchester, 2000); Aughton, Peter, *The Transit of Venus: the brief, brilliant life of Jeremiah Horrocks* (London, 2004).

[5] Howse, Derek, *Greenwich Time and the Longitude* (London, 1997), 35.

[6] According to J.L.E.Dreyer (*Flamsteed's letters to Richard Towneley*, The Observatory, v.45 (1922), 281), this could be a slip of the pen by Flamsteed, who probably meant Richard Towneley, not Christopher.

[7] Baily, Francis, *An Account of the Revd John Flamsteed* (London, 1835), 29–30.

Fig. 4.2 Christopher Towneley (Painted c.1630. © Towneley Hall Museum & Art Gallery)

> Amongst Mr. Gascoigne's letters I found some wherein he showed how the images of remote objects were formed in the distinct base of a convex object glass. From these I got my dioptrics in few hours; having read Descartes' Dioptrics before, but learnt little by them, because he discourses not of this subject.[8]

Filled with admiration by what he read, from that time onward he regularly credited Gascoigne with much of his own instruction in the science of telescopic optics. William Molyneux, in the preface to his *Dioptrica Nova* (1692), testified to the value of the material that Flamsteed had unearthed in the letters:

> I can say that the Geometrical method of Calculating a Rays Progress, which in many particulars is so amply delivered hereafter, is wholly new, and never before publish'd. And for the First Intimation thereof, I must acknowledg my self obliged to my worthy Friend Mr. Flamsteed, Astron. Reg., who had it from some unpublished Papers of Mr. Gascoignes.[9]

Flamsteed made efforts to get hold of as many remnants of the Gascoigne-Crabtree correspondence as he could. Later in the year, he wrote to John Collins,

> Sir, I understood by Mr Townlys Unkle [Christopher Towneley] that Mr Jonas Moore has severall of Mr Gascoignes and Mr Crabtrees papers or letters which having one parte of them in my hands I would gladly see because they often make reference to some others that I have not and which Mr Christopher Townly thinkes are in Mr Moores hands, but I know not how to move Mr Moore about them I am an egregious debtor to him already and such manuscripts I know to be a treasure of that nature as is not to be trusted to every one. and I can not tell how with civility to demand them of him, but you having given mee notice of these and some information of his I would entreat you when you have a convenient opportunity that you would please speake to him about them what your owne discretion may dictate to you so that they may come safe to my hands: I desire to know whether some partes or the summe of them have not beene excerpt by Dr Wallis into his collection of Mr Horroxes papers, for I find that there was a constant intercourse of mathematicall dissertations betwixt Mr Crabtree and Gascoigne and that it reacht to Mr Horrox too, tho Gascoigne and Crabtree became acquainted but some little while before Mr Horrox death: I esteem Mr Gascoigne by his papers to have beene as ingenuous a person as the world has bred or knowne yet better versed in Mechanik inventions and happier in them then in his Astronomy.[10]

The micrometer given by Jonas Moore was put to good use. Immediately, he started to check the exactitude of its scale: for every inch moved by the pointer, the screw made 35 full revolutions (measured on a linear scale) and 7 hundredths of a revolution (indicated by the index on the circular plate)[11]: In other words, it could divide a foot into more than 42,000 parts! Its performance in use was every bit as remarkable as the Royal Society had been lead to believe and Flamsteed sang its praises to Towneley:

> Had not the many curious and wonderful inventions of this age already prepared it to thinke nothing in nature not discoverable, I should very hardly, I beleive, have persuaded the world

[8] *Ibid*, 31–32.

[9] William Molyneux, *Dioptrica nova: A treatise of dioptricks* (London, 1692), Admonition to the Reader, 2.

[10] Forbes, E.G., Murdin, L., and Willmoth, F. (Eds), *The correspondence of John Flamsteed, The First Astronomer Royal*, v.1 (1666–1682), Letter to Collins, 1 August 1671, 102–3.

[11] *Ibid.*, v.1, Letter to Sir Jonas Moore, April 1674, 294–295.

Fig. 4.3 Towneley Hall (Author's collection)

> that Caelestiall observations may be made with that exactnesse I have daily experienced, by the helpe of your curious tho little engine applyed to a telescope[12]

He told Molyneux that 'the beginning of my observations at Derby for the first valewable Measures that I tooke in the heavens were got with that Micrometer'.[13] It is probable that it was used to make the important observations of Mars, which Flamsteed reported to Oldenburg on 10th November 1672:

> Last September I was at Towneley [Fig. 4.3] the first week that I intended to have observed Mars. There with Mr Towneley I twice observed him [Mars], but could not make two observations as I intended in one night. The first night after that of my return I had the good hap to measure his distances twice from two stars the same night: whereby I find that his parallax was very small certainly not 30 seconds: so that I believe the Sun's is not more than 10 seconds.[14]

It was in relation to exactly this type of measurement that the crucial importance of Gascoigne's invention was demonstrated. The parallax of the planets—and, linked to this, their distance—was a hot topic. And only the most delicate of instruments could suffice to reveal it.

---

[12] Ibid., v.1, Letter to Richard Towneley, 15 February 1674, 274–275.

[13] *Ibid.*, v.2, Letter to William Molyneux, 10 May 1690, 421.

[14] *Ibid.*, v.1, Letter to Henry Oldenburg, 16 November 1672, 185.

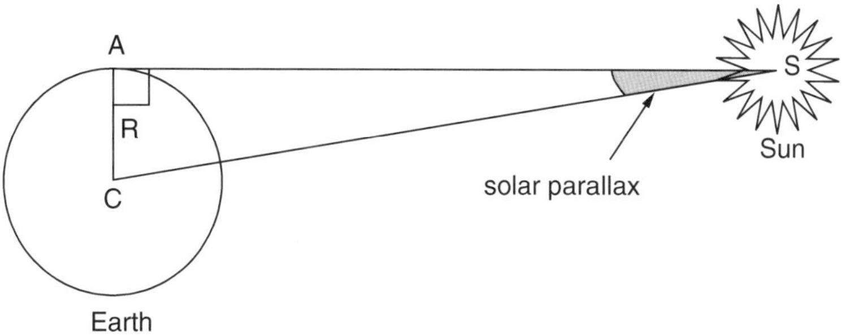

Fig. 4.4 The angle ASC is the solar parallax and CS = R/sin (solar parallax) (the Earth's size is greatly exaggerated relative to the Sun)

To appreciate the link between parallax and distance, let us imagine ourselves looking down on the Earth, from above the North Pole, at a time when the Earth is at its mean distance from the Sun. Then let us imagine an observer, on the Equator at A, seeing the Sun on the horizon, as shown below.

Our observer's line of sight to the Sun's centre is not in quite the same direction as the line from the centre of the Earth to the Sun's centre. This is the parallax effect. Looking at an object from a slightly different vantage point seems to shift its position. The effect can be clearly seen if we look at a nearby object—say a finger, held at arm's length—first with one eye and then with the other. The object appears to shift from side to side, relative to more distant things. Its direction differs according to which eye we use.

In the same way, the line of sight to the Sun from point A in Fig. 4.4 appears to converge with that from point C. The further away the Sun, compared with the radius of the Earth, the smaller the angle of convergence seems to be. This angle is called the *solar parallax*. It can also be described as *the angular semi-diameter of the Earth as seen from the Sun*.

Look again at the triangle formed by points ASC in the diagram. The length of the side AC is equal to the Earth's radius. The angle at A is a right angle—in other words, 90 °. If we can find out the size of the shaded angle, at S, the *solar parallax*, then, without leaving the Earth, we will be able to work out the length of side CS: the distance of the Sun—also known as one Astronomical Unit.

Unfortunately, such an exceedingly small angle as the solar parallax is difficult to measure. The difficulty is compounded by the nature of the Sun itself. In the simplest procedure, the parallax of a celestial body could be discerned as a slight angular shift when the body is observed from widely separated locations. With the Sun however, such is its brilliance, no background stars are visible in the sky, against which any such shift may be gauged. A more indirect route to the gauging of this parallax is needed—one which makes use of the planets and the laws governing their motion.

The Sun's parallax is today accepted as 8.794148″ (corresponding to a mean distance of 149,597,870 km). Flamsteed's figure of 10″, derived from using

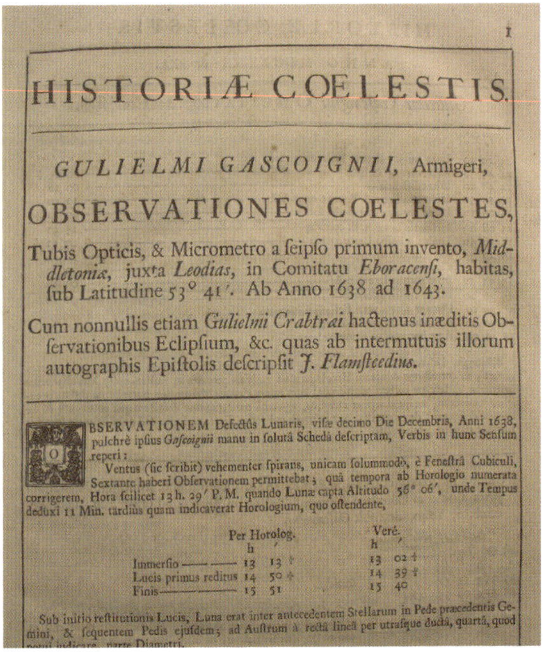

I

HISTORIÆ COELESTIS.

GULIELMI GASCOIGNII, Armigeri,

OBSERVATIONES COELESTES,

Tubis Opticis, & Micrometro a ſeipſo primum invento, *Middletoniæ*, juxta *Leodias*, in Comitatu *Eboracenſi*, habitas, ſub Latitudine 53° 41′. Ab Anno 1638 ad 1643.

Cum nonnullis etiam *Gulielmi Crabtræi* hactenus inæditis Obſervationibus Eclipſium, &c. quas ab intermutuis illorum autographis Epiſtolis deſcripſit *J. Flamſteedius.*

OBSERVATIONEM Defectûs Lunaris, viſæ decimo Die Decembris, Anni 1638, pulchrè ipſius *Gaſcoignii* manu in ſolutâ Schedâ deſcriptam, Verbis in hunc Senſum reperi:

Ventus (ſic ſcribit) vehementer ſpirans, unicam ſolummodò, è Feneſtrâ Cubiculi, Sextante haberi Obſervationem permittebat; quâ tempora ab Horologio numerata corrigerem, Hora ſcilicet 13 h. 29′ P.M. quando Lunæ capta Altitudo 56° 06′, unde Tempus deduxi 11 Min. tardiùs quàm indicaverat Horologium, quo oſtendente,

	Per Horolog. h ′	Verè. h ′
Immerſio ———	13 13 ½	13 02 ½
Lucis primus reditus	14 50 ½	14 39 ½
Finis ———	15 51	15 40

Sub initio reſtitutionis Lucis, Luna erat inter antecedentem Stellarum in Pede præcedentis Gemini, & ſequentem Pedis ejuſdem; ad Auſtrum à rectâ lineâ per utraſque ductâ, quartâ, quod potui indicare, parte Diametri.

Fig. 4.5 'Celestial observations of William Gascoigne' on the opening page of Flamsteed's *Historiae Coelestis* (1725) (Reproduced with the permission of Leeds University Library)

Gascoigne's micrometer was remarkably close: Perhaps too close, given the subtlety of the quantity and the fairly primitive nature of this early micrometer. Nevertheless, it was a pioneering measurement, which pointed the way for successful attempts (for instance, during David Gill's famous expedition to Ascension Island in 1877) in the late nineteenth century, by which stage telescope optics and the precision manufacture of micrometers had advanced considerably.

Flamsteed's tireless work in tracing fragments of the Gascoigne-Crabtree correspondence did not lead to immediate publication. In fact, it was not until 1725 that any material from these transcriptions was published. Some of Gascoigne's seminal observations of lunar, solar and planetary diameters were given pride of place on the opening pages of Flamsteed's magnum opus *Historiae coelestis Britannicae* (Fig. 4.5). The remaining contents of the transcribed correspondence[15] have never been published and have scarcely been referred to in any secondary literature.

[15] The National Archives, RGO 1/9, ff.3r-6r; 1/40, ff.9v-22r. Some of the content of the Crabtree-Gascoigne letter of 21 June 1642 was published in *The Gresham lectures of John Flamsteed*, by Eric G.Forbes (London, 1975), 51–2.

Chapter 5
Derham and De La Hire

Following Sherburne's *Astronomical Appendix*, more than 30 years would elapse before anything more concerning William Gascoigne appeared in print. It came thanks to the exertions of the Reverend William Derham (1657–1735)—Anglican rector of Upminster and member of the Royal Society (Fig. 5.1). Astronomy was one of Derham's great passions, but like many early members of the Society he was also drawn to all manner of topics concerning the natural world. He collected bird, insect and botanical specimens; kept registers of atmospheric pressure and temperature, and made one of the earliest worthwhile measurements of the velocity of sound. He also made prolific observations of sunspots and the planet Jupiter.

To Derham the whole world teemed with the wonders of God's creation, just waiting to be studied. Scientific enquiry for him was not a threat to religious faith: Just the opposite.

> Let us…inspect every Part…search out the inmost Secrets of any Creatures; *he wrote*, let us examine them with all our Gauges, measure them with our nicest Rules, pry into them with our Microscopes, and most exquisite Instruments, still we find them to bear Testimony to their infinite Workman.[1]

His book *Physico-Theology: or, A Demonstration of the Being and Attributes of God from his works of Creation* (1713) exemplified this conviction and earned him admirers in many quarters. Samuel Johnson's friend Hester Thrale demanded to know, 'What Christian lives who can refuse his reverence to Derham's learning?[2]' Dr. Johnson himself, on his Hebridean Tour confided to his companion, James Boswell, 'I do not like to read anything on Sunday, but what is theological' and how on the Sabbath, for this reason, he 'then took up Derham's *Physico-Theology*'.[3]

---

[1] Derham, William, *Physico-Theology: or, a Demonstration of the Being and Attributes of God from His Works of Creation*, 5th edition (London, 1720), Book II, 37–8.

[2] Atkinson, A.D., William Derham, FRS (1657–1735), Annals of Science, v8 (1952), 368–9.

[3] Rogers, Pat (Ed.), *Johnson & Boswell in Scotland. A Journey to the Hebrides*, New Haven & London, 1993, 256.

D. Sellers, *In Search of William Gascoigne: Seventeenth Century Astronomer*, Astrophysics and Space Science Library 390, DOI 10.1007/978-1-4614-4097-0_5,

Fig. 5.1 William Derham (From *Science in Oxford*, xi, by R.T. Gunther, 1937)

Building on the same theme, of what we would now call 'intelligent design', Derham quickly went on to prepare another work, which concentrated on astronomical topics: *Astro-Theology: or, A Demonstration of the Being and Attributes of God, from a survey of the Heavens* (1715).

His arguments were perhaps not seriously challenged until David Hume published his *Dialogues concerning natural Religion*, in 1779.

But it was the study of rainfall that brought Derham, regardless of different religious affinities, into close friendship with Richard Towneley. In order to examine the variability of rainfall across the country, the two men had engaged in an extensive collaboration—both meticulously measuring and comparing daily rainfall statistics in their respective localities.

Maybe Towneley, aware of Derham's astronomical activities, had discussed the work of Gascoigne with him. At any rate, Derham's interests were known to the wider Towneley family. Consequently, when Richard Towneley's 'scatter'd Papers and Letters' of Gascoigne eventually passed into the hands of his son, Charles Towneley (1658–1712), he, in the year before his death, gave them to Derham.

Derham deemed the papers from Richard's son to include 'most of the Letters between Mr. Crabtrie and Mr. Gascoigne' that the Towneleys had recovered. By the Spring of 1711 he had already published one of them in the *Philosophical transactions*: a lengthy letter from Crabtree to Gascoigne (7 August 1640) mainly about the nature of sunspots.[4]

This was the first letter from Crabtree to Gascoigne and pre-dates any meeting between the two. It appears that Richard's uncle, Christopher Towneley (or one of his acquaintances, called Kay) had written to both Crabtree and Gascoigne asking for their opinions on the nature of sunspots. The responses were contradictory. Crabtree, taking a position similar to the Italian scientist Galileo, considered that the spots were blemishes on the face of the Sun and their movement from east to west was due to the rotation of the Sun itself. Gascoigne, by contrast, erroneously thought that the spots were actually the silhouettes of small planets locked into a close orbit around the Sun. On learning of the divergence of opinion, Gascoigne wrote to Crabtree giving a detailed account of his reasons. The two had been unacquainted up to this point, but Crabtree, sensing a kindred inquiring spirit, wasted no time in sending a reply—signing himself 'Your Loving Friend (though *de facie ignotus*[5])'.

Despite the opposing conclusions on the spots, in his letter Crabtree expresses his appreciation of the originality of Gascoigne's reasoning and preference for relying on observational proofs rather than authority. The insistence on first-hand observation precisely matches his own persuasions:

> I do not value the Authority of Galilaeus (though reputed the greatest Speculative Mathematician in Europe) nor yet Kepler (though *Astronomorum facile princeps*[6])

[4] *Philosophical transactions*, xxvii (1712), 280–289.

[5] *de facie ignotus*: of unknown countenance.

[6] Easily the prince of astronomers.

further than either Demonstrative, or the most probable Reasons confirm their Opinions. Nor will I stick to subscribe to the Man whosoever shall bring better Reasons for his Opinion.[7]

In his letter Crabtree shows a keen interest in the novel optical discoveries and instruments that Gascoigne has apparently made. It does not appear that he is as yet aware of much detail concerning these and is not certain how open Gascoigne is willing to be on this score.

...let me encourage you to proceed in your noble Optical Speculations. I do believe there are as rare Inventions as *Galilaeus Telescope*, yet undiscover'd. My living in a place void of apt Materials for that purpose, makes me almost Ignorant in those Secrets; only what I have from Reason, or the reading of *Kepler's Astron. Opt.*[8] and *Galilaeus*. If you impart unto us any of your Optical Secrets, we shall be thankful, and obliged to you, and ready to requite you in any thing we can.[9]

He gives an account of the observation of the 1639 transit of Venus made by himself and Horrocks and promises to bring Jeremiah Horrocks—'my second Self'—and [Christopher] Towneley to visit Gascoigne in Yorkshire, along with observations of the most recent solar eclipse, made by himself in Broughton, Horrocks in Hoole and Samuel Foster in London. Although Crabtree expects his new friend from Yorkshire to be familiar with the works of Ptolemy, Tycho, Galileo, Kepler and Lansberge, he is evidently unsure as to whether Gascoigne embraces keplerian elliptical orbits and takes some care to explain why they alone adequately describe the motion of the planets.

Kepler's Elliptick is undoubtedly the way which the Planets describe in their Motions: And if you have read his *Comment. de motu Martis*, and his *Epit. Astron. Copern.*[10] I doubt not you will say his Theory is the most rational, demonstrative, harmonious, simple, and natural that is yet thought of, (or I suppose can be;)[11]

As for the other Gascoigne-Crabtree letters, Derham complained that 'being long, I have not time at present to fit them up for the Societies Use; but intend (God willing) to do it as soon as may be, if this Specimen be acceptable'.[12]

Within the English scientific community, the matter of Gascoigne's priority had long since been settled by the publication of Towneley's account in the *Philosophical transactions*. Even so, Justel's impersonal reference to Gascoigne simply as 'the person who is dead', was perhaps a sign that Gascoigne's name would not easily make its mark outside his native country. Consequently, when a generation later Philippe de La Hire (1640–1718) mentioned the telescopic sight in the introduction to his astronomical tables of 1687, it was almost inevitable that he should overlook

[7] *Philosophical transactions*, xxvii (1712) 280.

[8] *The Optical Part of Astronomy* (1604).

[9] *Philosophical transactions*, xxvii (1712) 287.

[10] *New Astronomy* (1609) and *Epitome of Copernican Astronomy* (1618–1621).

[11] *Philosophical transactions*, xxvii (1712) 288–9.

[12] *Ibid.*, 290.

Fig. 5.2 Philippe de La Hire (Author's collection)

the young English astronomer and reiterate the presumed priority of the French—citing Picard in particular.

> A few years ago, D.Picard, the distinguished astronomer, and a member of the same Academy [Royale des Sciences] removed sighting notches from instruments and in their place substituted telescopes; whereby distant and nearby objects together remained in sharpest vision.[13]

De La Hire (Fig. 5.2) was an accomplished Parisian geometer and had produced work of lasting value on the theory of conic sections. After entering the prestigious Académie Royale des Sciences, he had been assigned by Colbert—Controller-General of Finances for Louis XIV—to assist Picard with the surveys and observations which eventually led to a new Atlas of France (the very one which, showing France to be smaller than hitherto thought, prompted the King to protest that he had lost more land at the hands of his astronomers than he had ever forfeited to his enemies in war). It was only to be expected that de La Hire should be most familiar with the role of Picard in pioneering the use of the telescopic sight.

The telescopic sight, of course, although not as complex as the micrometer, utilised the same principle of a simple marker at the common focal point within the telescope. At this position the marker—be it a wire or a strand of silk—could serve to identify the precise position of a distant object within the field of view. Attached to a land-surveyor's instrument, such as the theodolite and its predecessors, an accessory of this character would enable a step-change in the accuracy with which angular separations could be measured when carrying out triangulation surveys.

The same applied in astronomy too. Gascoigne's first trial of the principle, more than two decades before the work of Auzout and Picard, had also involved the introduction of a simple marker into a small telescope, thus creating a telescopic sight. By this means he could time the passage of a celestial body over a specific point

[13] *Philosophical Transactions*, xxx (1717), 603 (and extended from the Preface of *Tabularum Astronomicarum* of Philippe de la Hire, published in Paris in 1687).

within the field of view. By attaching a small telescopic sight to his quadrant he could achieve much more accurate alignments. Shortly afterwards he converted the marker into a tiny graduated scale for direct measurement of angular sizes or separations. Then, with an ingenious imaginative leap, he devised the micrometer—the instrument rescued by Towneley—by replacing the tiny scale with a pair of screw-controlled moveable pointers, whose mutual separation could be read off an external scale.

On seeing de La Hire's words, Derham was just as offended as Richard Towneley had been 40 years earlier. It was not surprising that Derham's feelings echoed the earlier ones of Towneley: As the new custodian of the few remaining Gascoigne-Crabtree papers, it was now he who must protect the record that they contained of the endeavours of their authors.

On St. Andrew's Day (30 November) in 1713, spurred on by a desire to defend Gascoigne, Derham was moved to write to Richard Waller, then secretary of the Royal Society, in the following terms:

> Had I seen you I intended to have desired you to acquaint the Soc. next Thursday, That Monsr de la Hire in the 2nd Page of his Preface to the first part of his Astron. Tables, robs our Mr Gascoigne of the hon^{r} of first applying Telescopick Sights to mathematical Instrumts, and ascribes it to Monsr Picard, a few years before the year 1686. Now since Mr Gascoignes Papers are in my hands, and I am fully able to do Mr Gascoigne justice, if the Soc pleases I will send you copies of some of that admirable Gentleman's letters of 1638 &c to be inserted in your Transactions (if you intend to print any more of them).[14]

By July 1715 Derham had prepared much of the promised paper and on 14 Jul 1715 wrote to the Leeds antiquary, Ralph Thoresby (1677–1724), seeking any biographical information about Gascoigne that might serve to supplement the paper.[15] In the plea to Thoresby, Derham mistakenly characterised Gascoigne as 'an admirable son of Sir William Gascoigne' who was 'killed at Marston Moor battle, at 23 years of age'. As it happened, Thoresby was about to publish his monumental *Ducatus Leodiensis*,[16] and had included therein an extensive pedigree for the wider Gascoigne family. In this pedigree he had taken care to include the Gascoignes of Thorp-on-the-Hill (a hamlet near Middleton) 'for the sake of William Gascoigne, Esq; the last of that Branch, who is deservedly celebrated for his Astronomical Discoveries and Mathematical Genius'.[17] The pedigree showed William to be the eldest son, not of Sir William Gascoigne, but of Henry Gascoigne and Jane Cartwright. It was not revealed when he was born, but his age when slain in 1644 is implicitly shown to be not as claimed by Derham (23), since his mother is stated to

[14] Royal Society archive, *EL/D1/60*. Letter of Derham to Waller (30 November 1713).

[15] *Letters of eminent men addressed to Ralph Thoresby, FRS* (London, 1832), ii, 302.

[16] Thoresby, Ralph, *Ducatus leodiensis: or, The topography of the ancient and populous town and parish of Leedes, and parts adjacent in the West-riding of the county of York*, etc, (London, 1715).

[17] *Ibid.*, 181.

have died in 1617. According to Thoresby, Gascoigne was killed at Melton Mowbray, rather than Marston Moor.

Derham's paper—entitled *Extracts from Mr. Gascoigne's and Mr. Crabtree's Letters, proving Mr. Gascoigne to have been the Inventor of the Telescopick Sights of Mathematical Instruments, and not the French*—finally appeared in the *Philosophical transactions* in the Spring of 1717. It concurred with Thoresby about the identity of Gascoigne's father, but ignored the inconsistencies relating to his age.

Derham explained that he now had all of the Gascoigne papers that had been rescued by Towneley and proceeded to give extracts from six letters: two from Gascoigne to Crabtree (25 January 1641, 24 December 1641) and four from Crabtree to Gascoigne (one undated, 30 October 1640, 28 December 1640, 6 December 1641). The extracts were selected to furnish incontrovertible evidence that Gascoigne was making and using telescopic sights and micrometers in the early 1640s. They succeeded amply in this task, but afforded little in the way of additional information about the life and circumstances of their authors.

It is evident from the letter that Crabtree dispatched to his new friend on 30 October 1640 that he had made an impromptu visit to Middleton, perhaps on the way back from a business trip, since receiving the first letter from Gascoigne. Whilst there, he had been shown a host of unfamiliar instruments along with explanations of the principles by which they operated. Too much for him to absorb in the short time available.

> I not a little lamented that my Time cut me so short, when I was with you, that I could not more fully ruminate and digest those strange Inventions which you showed me, and told me of. My Lassitude after an unexpected and unacquainted Journey; my unpreparedness for those Cogitations (not intending that Journey the Day before) and the Multiplicity and Variety of the Novelties you shewed me, so wholly distracted my Thoughts into Admiration, that I cannot now give my Meditations any reasonable Account of what I saw: but must intreat you, in a few Lines, to rub up my Memory, and tell me again what you shewed me, and the Extent of those your Inventions....
>
> You told me (as I remember) you doubted not in time to be able to make Observations to Seconds. I cannot but admire it and yet, by what I saw, believe it: but long to have some farther Hints of your Conceit for that Purpose. One Means, I think, you told me was, by a single Glass in a Cane, upon the Index of your Sextant, by which (as I remember) you find the exact Point of the Sun's Rays. But the way how, I have quite forgotten, and much desire. Your Device for the exact Division of a Quadrant, by dividing 11 Degrees into 10 Parts, I did then understand, but do not now fully remember. If it might not be too much Trouble to you, I should intreat you to give me such a Paper-Demonstration thereof as you shewed me, and two or three Lines plainly of the Use thereof, how to find those small Parts. I lost the little Paper, wherein I noted the Moon's Diameter, which we observed when I was with you: I pray you send it me, if, &c.[18]

Although Crabtree had failed to remember much technical detail, he is certain that he had been shown priceless additions to the arsenal of astronomy that left him full of yearning.

[18] *Philosophical Transactions*, xxx (1717), 606–7.

I cannot conceal how much I am transported beyond my self with the Remembrance (of that little I do remember) of those admirable Inventions which you shewed me when I was with you. I should not have believed the World could have afforded such exquisite Rarities, and I know not how to stint my longing Desires, without some further Taste of these selected Dainties. Happier had I been, had I never known there had been such Secrets, than to know no more, but only that there are such ....

.... although Modesty would forbid me to request any thing (until you give me leave) but what you please voluntarily to impart, yet the Vehemence of my Desire forceth me to let you know how much I desire, and how highly I should prize any thing that you should be pleased to communicate to me in those Optick Practices. Could I purchase it with Travel, or procure it with Gold, I would not long be without a Telescope for observing small Angles in the Heavens; nor want the Use of your other Device of a Glass in a Cane upon the moveable Ruler of your Sextant (as I remember) for helping to the exact Point of the Sun's Rays.[19]

Crabtree soon passed the news these inventions on to Horrocks and it would be quite understating the matter to say that he too was impressed, as Crabtree made clear to Gascoigne in the next letter quoted by Derham (dated 28 December 1640).

My Friend Mr.Horrox professeth, that little Touch which I gave him of your Inventions, hath ravished his Mind quite from it self, and left him in an Extasie between Admiration and Amazement. I beseech you, Sir, slack not your Intention for the perfecting of your begun Wonders. We travel with Desire till we hear of your full Delivery. You have our Votes, our Hearts, and our Hands should not be wanting, if we could further you.[20]

In response to the earlier request for further details it is apparent that Gascoigne in due course has supplied diagrams and a mathematical analysis, for Crabtree expresses his admiration of them.

Your Diagrams for Perspectives [telescopes] I have viewed again and again, and cannot sufficiently admire your indefatigable Industry, and profound Ingenuity therein. I am much affected with the Symbolical Expressions of your Demonstrations. I never used them before (but I will do) yet I understand them all at the first Sight, and see well the Truth of your Demonstrations.[21]

Derham gives only the briefest of extracts from the lengthy letter of Gascoigne to Crabtree, which is dated 25 January 1641, but this provides ample confirmation that Gascoigne has invented both a rudimentary micrometer and a telescopic sight. His advice about how to illuminate the sighting 'hairs' shows that he is by now making practical use of the devices.

If here [that is in the distinct base] you place the Scale that measures—, or if here an Hair be set, that it appear perfectly through the Glass—, you may use it in a Quadrant, for the finding of the Altitude of the least Star visible by the Perspective wherein it is. If the Night be so dark, that the Hair or the Pointers of the Scale be not to be seen, I place a Candle in a Lanthern, so as it cast Light sufficient into the Glass; which I find very helpful when the Moon appeareth not, or it is not otherwise light enough.[22]

---

[19] *Ibid.*, 607–8.

[20] *Ibid.*, 608.

[21] *Ibid.*, 608.

[22] *Ibid.*, 604.

By the following December it is apparent that the micrometer has progressed from being a simple scale, inserted within the telescope tube, to being a fully-fledged instrument that uses sliding pointers whose separation is determined by a couple of screws. Crabtree's letter of 6 December 1641 summarises its unique potential.

> That which you give me a full Projection of was above my Hope: and if the Screws keep an exact Equality of Motion forward in each Revolve, it is a most admirable Invention; and with the other Accomodation, I had almost said without Compare. But that the Divisions of a Circle should be measured to Seconds, without the Limb of an Instrument, or that Distances, Altitudes, Inclinations, and Azimuths should be taken all at one Moment, without the Limb of an Instrument likewise, and each to any required Number of Parts; or that the Diameter of Jupiter should be projected in such prodigious Measures as you speak of, &c. were enough to amuse and amaze all the Mathematicians in Europe, and may indeed be rather a Subject of Admiration than Belief, to any that hath not known your former Inventions to exceed Vulgar (I had almost said Humane) Abilities. And for my Part, I must confess Modesty so checks my ambitious Desires, that I dare scarce hope such Miracles should ever be produced in real Practice to such Exactness.[23]

Tragically, the chance of Gascoigne meeting Horrocks no longer exists as the latter had died, barely 22 years old, of unknown cause earlier in the year and Crabtree is still feeling the loss.

> No Man that hath written of the Diagram [of Hipparchus] understood it fully or described it rightly, but only Kepler and our Horrox; for whose immature Death there is scarce a Day which I pass without some Pang of Sorrow.[24]

Nevertheless, Gascoigne was still intent on further modifications to his instruments in order to test Horrocks' lunar theory and a further (undated) letter shows his ambitions.

> I have given order for an Iron Quadrant of Five Foot, which will give me the 1000th Part of One Degree, which shall be furnished like my first Scale; only my Workman is so throng for my Father, that I fear it will not be finished before the Eclipse. I have caused a very strong Ruler to be exactly made, and intend to fit it with Cursors of Iron, with Glasses in them and a Thread, for my Sextant.[25]

This culminates in the final letter to Crabtree (24 December 1641) quoted by Derham.

> Mr Horrox his Theory of the Moon I shall be shortly furnished to try. For I am fitting my Sextant for all manner of Observations, by two perspicills with Threads. And also I am consulting my Workman about the making of Wheels like β, γ, δ, ε, of † Diagr.3, to use two Glasses like a Sector. If I once have my Tools in readiness to my Desire, I shall use them every Night. I have fitted my Sextant by the Help of the Cane, two Glasses in it, and a Thread so as to be a pleasant Instrument, could Wood and a Country-Joiner please me.[26]

---

[23] *Ibid.*, 608–9.

[24] *Ibid.*, 609.

[25] *Ibid.*, 605.

[26] *Ibid.*, 605.

Satisfied that he had done justice to the memory of Gascoigne, Derham concluded his paper.

> Thus, among many, I have related some of the Passages of Mr Gascoigne's and Mr Crabtrie's Letters relating to Telescopick Sights. From whence it is very manifest, that long before the French Gentleman's Claims, our Countryman Mr Gascoigne had made use of those Sights in his Astronomical Instruments; particularly in two or more Sorts of Micrometers (as I plainly find) and in his Quadrant and Sextant. And had it pleased God to have given him a longer Life, we might have expected greater things from his pregnant and sagacious Wit.[27]

It is a great pity that Derham published only the extracts from a few chosen letters, since—following his death—this body of manuscripts disappeared. The Oxford historian Allan Chapman echoed the views of many scholars when he wrote that this material was 'a correspondence which, while surviving in at least ten detailed letters in 1717, has subsequently vanished, so that all we have are the selections which Derham edited for the Royal Society.'[28] Nevertheless, the situation is not quite as bleak as this, since extensive copies of the Gascoigne-Crabtree correspondence are available elsewhere, as we shall see.

The Reverend William Derham died on 5th April 1735 at the age of 77. He was a modest, unassuming man to the end and his will requested that he 'receive Christian burial in a decent but withal frugal manner'. Consequently there is no memorial to him in the Church of St Laurence in Upminster that harbours his remains. To his widow, Anna, he left the bulk of his property, though a special provision was made for his 'philosophical' possessions.

> … imagining that my Books, Telescopes and other Mathematical and Philosophical Instruments are best understood and will be of greater use to my dear Son William Derham than to any other of my Children I therefore bequeath to him all my Books Telescopes and Telescopick Glasses of all the several Lengths and all other Mathematical and philosophical Instruments, Manuscripts, and Papers, excepting such Books as my dear Wife may think fit to reserve for her own use.[29]

The manuscripts passed on to William Derham junior presumably included the Gascoigne-Crabtree letters.

Like his father, William Derham junior took holy orders—becoming the Reverend Doctor Derham. He was soon a respected figure in academic circles and in due course rose to become President of St John's College in the University of Oxford. Unfortunately, he did not prove to have the longevity of his parents and died in 1757 at the age of 54. A few months beforehand, sensing the proximity of death, he dictated a will that left to his 'honoured Mother, Mrs Anna Derham … all my Estate of what Nature soever whether in Money, Goods, Chattels, Books, Household or Furniture'.[30]

---

[27] *Ibid.*, 609–10.

[28] Chapman, Allan, *Jeremiah Horrocks and the transit of Venus of 1639*, in Transits of Venus: New views of the Solar System and Galaxy (IAU Colloquium 196), (Cambridge, 2004), 12.

[29] The National Archives, Public Records Office, Prob/11/672.

[30] The National Archives, Public Records Office, Prob/11/833.

Thus, assuming that the Gascoigne-Crabtree letters had not previously been disposed of, his mother became their new custodian. Anna quickly made arrangements to pass on the collection of 'philosophical' material to William junior's cousin, her nephew, George Scott FRS (1719–1780) of Woolston Hall, Chigwell, in Essex.

Scott was taken aback by the wondrous range of the material. He wrote to his brother on 30th August 1757,

> I have just left off looking at a large collection of manuscript letters from some of the greatest men of our Age…by the Death of Dr Derham [junior] I am possessed of a very large and curious Collection of Books, Manuscripts, Pictures, Prints, Shells, Medalls, Coins, Drawings, Fossils, Seeds, mathematical Instruments, Seals, neat Tolls of all kinds, etc. etc. etc. In short I am astonished at the learning and Industry of the late President of St John's, tho' I knew him so well; and it gives me great Pleasure that so good and great a Man should express so high a value for me to my Aunt and Cousin in his Life Time, and take such Pains to collect and methodize his Collection for me. I worked like an Horse for eight days at the Lodgings, and had not my Relation been so very exact, as many months would not have been time enough for me to have gone through his Collection in, and hope to have packed up the thousands of Articles now here.[31]

William junior had indeed been a very dear cousin and George Scott lamented that he had no portrait to remember his learned cousin by. In the same letter he advised his brother, 'You do very right to have a Picture of your Son; I would almost give one of my Ears to have one of my deceased Relation; he in his Life Time would not hear of any such thing, and now as we say in Oxon, *Tempus preterlapsum*'.

Although the Oxford bookseller Daniel Prince advertised the sale on 18th October 1757 of the library of books belonging to William Derham, the sale catalogue didn't include mention of any manuscripts.[32] There was talk of giving the manuscripts to St. John's College, but Scott advised one of his cousins 'Do not take any other Opinion than what you have in your own Family, about giving up the late good President's manuscripts to the College, which they ought not to have till they deserve better than they do at present to possess such a Treasure'.[33] The only manuscripts of Derham provenance to be found in the College library now are a few leaves of lecture notes by William senior. We must assume that, in the family's estimation, the College never did become deserving!

When George Scott died without children in 1780—a virtual recluse at Woolston Hall—his estate, it is thought, passed to his cousin and on 12th March 1781 a grand 16-day auction began. The Covent Garden booksellers, Leigh and Sotheby, were charged with disposing of what they described as 'the Entire and Valuable Library of

[31] Bodleian Library, MS.Eng.e.3655, George Scott Letter book, v.35, 23–25. A typed transcription also exists in Essex Record Office, D/DU 546/3.

[32] Prince, Daniel (bookseller), *A catalogue of books; being the library of…William Derham. Which will begin to be sold Oct.18, 1757* (Oxford, 1757).

[33] Bodleian Library, MS.Eng.e.3655, George Scott Letter book, v.35, 50–51. A typed transcription also exists in Essex Record Office, D/DU 546/3.

George Scott, Esq, … containing a fine collection of books…Also his collection of manuscripts'. By the end of the sale 3,457 lots had been dealt with. Although some Derham material was clearly included in the sale, the only lots which could have contained the Gascoigne-Crabtree material were anonymously described simply as 'A Large Parcel of MSS [manuscripts] upon different subjects; A Parcel upon various Subjects; and, A large Parcel upon different Subjects, containing original Letters …' (no Derham provenance being indicated).[34] We do know that part of the collection also remained within his family and eventually passed, along with Woolston Hall itself, to the Perry Watlington family of Moor Hall at Harlow in Essex, who were responsible for the sale of some manuscripts in 1891. It was at this sale of items from the Moor Hall library that the Guildhall Library acquired invaluable papers of the scientist Robert Hooke, including an extensive Memorandum Book.

Amid all the confusion of legacies, bequests and other transactions, the fate of the Gascoigne-Crabtree correspondence—the letters that Jonas Moore, the Towneleys and William Derham had so carefully striven to preserve—is completely unknown.

[34] Bodleian Library, Vet.A5.e.7132, *Sale Catalogue of George Scott*, sixteenth day's sale (Manuscripts, Octavo).

Chapter 6
Bevis and De La Hire

At almost the same time as William Derham's account of the Gascoigne priority was being printed in the *Philosophical transactions*, de la Hire was busy compounding his earlier 'misdemeanour'. His memoir of 29 May 1717 to the Académie Royale des Sciences concerning the construction and use of the Micrometer did not even mention Gascoigne.

> No one can doubt that the Micrometer is one of the most useful instruments in the practice of Astronomy. Its perfected construction is due to Messrs Auzout and Picard, as one can see in a printed journal, entitled *'Extract of a letter from M.Auzout of 28 December 1666 to M.Oldenburg, Secretary of the Royal Society of England, touching on the diameters of the planets, etc'*. And as this document is quite rare, I believe it is relevant, in order to preserve the memory of its invention, & for the honour of those who invented it, to have it reprinted in the posthumous works of Members of the Academy.[1]

A short time later, on 23 June 1717, he presented a further paper, going into the matter of priority. Here he explained that though Auzout and Picard *perfected* the instrument, even they had not invented it. Still Gascoigne was not mentioned.

> But, as regards the micrometer, near the end of M. Huygens book on *The system of Saturn*, printed in 1659, that is to say three years before the *Ephemerides* of the Marquis Malvasia went to press, we find the method of observing the diameters of the planets, using the telescope, and placing, as he says, at the convex eyepiece focus, which is also the focus of the objective, an object that he calls virgula, with a size suitable for encompassing the object that he might wish to measure, because he points out that at this place within the telescope,

[1] *Histoire de l'Académie Royale des Sciences (Année MDCCXVII), avec les Mémoires, etc* (Paris, 1717), 57.

D. Sellers, *In Search of William Gascoigne: Seventeenth Century Astronomer*, Astrophysics and Space Science Library 390, DOI 10.1007/978-1-4614-4097-0_6,

> with two convex glasses, one can see very distinctly the smallest objects, & it was by this means that he measured the diameters of the planets, as he reports them after having determined, from the experience of the passage of a star behind this body how many seconds of a degree it subtended. There is so little difference between the construction of the micrometer that M.Huygens made use of and that of the Marquis Malvasia, which only appeared three years later, that the latter cannot qualify as a discovery. Thus we have to concede that we are indebted to M. Huygens for the invention of the micrometer, and this was later perfected to its present state.[2]

Derham's reaction to de la Hire's tables, a full 20 years after their publication, was a model of alacrity compared with the response of Dr. John Bevis (1695–1771) to de la Hire's 1717 memoir. In a letter to James Short, which was read to a Royal Society gathering on 17 May 1753, and published in the *Philosophical transactions* later that year, Bevis complained that,

> No sooner had the late Dr Derham restor'd the application of telescopic sights to quadrants to its true author Mr.Gascoigne, than M. de la Hire, who never made the doctor any reply on that head, took occasion, in the memoirs of the Royal Academy of Sciences for 1717, to ascribe this contrivance of the micrometer to M.Auzout, in conjunction with M.Picard; alleging, for proof, an extract of a letter, dated Dec.28, 1666, from M.Auzout to M.Oldenburg ....
>
> I have now before me the copy of a letter of Mr.Gascoigne to Mr.Oughtred, which I made myself from the original, written in 1640–1; which original was in the possession of the late William Jones, Esq, F.R.S. and is now in the library of the right honourable the Earl of Macclesfield. It consists of several sheets of paper, all about his invention for measuring small angles to seconds; where he not only gives the geometrical and optical principle of his contrivance, and the construction of the instrument, but also a series of observations actually taken therewith; some of which I shall transcribe.[3]

Bevis went on to give some of the measurements of the angular diameters of Jupiter, Mars, Venus and the Moon that Gascoigne had included in his letter to Oughtred.

This letter, whose existence was revealed by Bevis, would later prove to be a source of valuable biographical information concerning Gascoigne, but for the time being it remained hidden in Macclesfield's library.

The *Journal Britannique* of the London-based Huguenot refugee, Matthieu Maty (1718–1776), was established in 1750 with the object of keeping the francophone world abreast of current English thought and new publications. It faithfully relayed the content of Bevis's letter for the benefit of continental philosophers. For the first time maybe, Gascoigne's name appeared in a French report.[4]

The next snippets of biographical material to be published came in the notoriously unreliable *Brief Lives* of Oxford polymath John Aubrey (1626–1697), the first version of which was published in 1813 (Fig. 6.1).

---

[2]de la Hire, P., *Recherche des dates de l'invention du micrometre, des horloges à pendule, & des lunettes d'approche*, Mémoires de l'Académie Royale des Sciences (1717), 80.

[3]*Philosophical transactions*, xlviii (1754), 190–191. The report that de la Hire's latest memoir (1717) credits Auzout and Picard with the invention seems at odds with his actual statement that Huygens was responsible.

[4]Maty, M., *Journal Britannique*, xvi, Jan-Feb 1755, 109–110.

Fig. 6.1 John Aubrey (From *Brief Lives*, ed. A. Clark, 1898)

> There was a most gallant gentleman and excellent mathematician that dyed in the late warres, one Mr. Gascoigne, of good estate in Yorkshire, to whom Sir Jonas Moore acknowledged to have received most of his knowledge. He was bred up by the Jesuites. I thought to have taken memoires of him; but deferring it, death took away Sir Jonas. But I will sette downe what I remember. Gascoigne, esq., of Middleton, neer Leeds, Yorkshire, was killed at the battaile of Marston-moore, about the age of 24 or 25 at most. Mr Edmund Flamsteed, who sayes he found out the way of improveing telescopes before Des Cartes, tells me, Sept.1682, that 'twas at Yorke fight he was slaine.[5]

Although the information set out by Aubrey subsequently found its way into other accounts, part of it—for example, regarding the age of Gascoigne—is clearly false. Even the forename of Flamsteed is incorrect: perhaps a slip of the pen, confusing Flamsteed's name with that of his erstwhile friend Edmond Halley. The brief statement that Gascoigne 'was bred up by the Jesuites' seems to have become the cornerstone for a belief by many later authors that Gascoigne was a Roman Catholic. There does not appear to be any documented evidence from other sources for this conviction.

In the section of *Brief Lives* on Sir Jonas Moore, Aubrey gave a surprising additional piece of information

> I remember Sir Jonas told us that a Jesuite (I think it was Grenbergerus, of the Roman College) found out a way of flying, and that he made a youth performe it. Mr Gascoigne taught an Irish boy the way, and he flew over a river in Lancashire (or therabout), but when he was up in the ayre, the people gave a shoute, wherat the boy being frighted, he fell downe on the other side of the river and broke his legges, and when he came to himselfe, he sayd that he thought the people had seen some strange apparition, which fancy amazed him. This was anno 1635 ...[6]

Far-fetched though it sounds, Moore had indeed given this story to a meeting of the Royal Society on 8th May 1679. Aubrey's account is inaccurate in only one respect: the site of this incredible feat was across the river gorge at Knaresborough,

---

[5]Aubrey, John, *Brief lives and other selected writings,* ed. Anthony Powell (London, 1949), 138.

[6]*Ibid.*, 148.

near Leeds, not in Lancashire.[7] As frequently happened at the Society's meetings, discussion had been sparked off by a mechanical model made by Robert Hooke. This purported to demonstrate with pasteboard wings a way of flying that Hooke claimed had been 'invented and practised by one Mons. Besnier, a smith of Sable in the county of Mayne, the contrivance of which consisted in ordering four wings folding and shutting'. By this method, powered by his own limbs, it was said that Besnier had been able to 'fly from a high place cross a river to a pretty distance'. Dr. Croone added that the Paris Gazette had recently reported someone flying safely from the top of a steeple some considerable distance. In response Jonas Moore related Gascoigne's experiment 'above 40 years before'.

News of these early attempts at aviation spread far and wide and intense discussion took place about the weakness of man's arms compared with the strength of birds' wings. Natural philosophers wondered whether they were hearing about real attempts at flight (was that even possible?), or a straightforward hoax: a magician's illusion.

The English mathematician, John Pell (1611–1685) mentioned the stories to the French monk Marin Mersenne (1588–1648) sometime in late 1639 or early 1640.[8] Mersenne acted as a kind of 'clearing house' for the interchange of news and ideas amongst the scientific circles of western Europe at this time and he in turn sought the opinion of others on the matter of human flight. To Professor André Rivet, a theologist in the Hague, he wrote alluding to 'strange things that certain engineers promise here, that I don't believe to be possible by simply natural means'.[9] In a similar vein he sought the view of René Descartes, who initially reserved his position, saying: 'I would like to see the effects before believing these propositions. It is possible to keep a body in the air for a short time, but not to make it stay there … the birds beat the air with their wings when they need to go forward or stop, something which cannot be imitated by machines made with human hands'.[10] A month later, still sceptical, but clearly intrigued by the idea, he advised:

> One could well make a machine which keeps itself in the air like a bird, *metaphysically speaking*; because the birds themselves, at least in my view, are just such machines; but not *physically* or *morally speaking*, because it would require 'springs' so light and yet so strong, that they couldn't be made by humans.[11]

By the early nineteenth century Gascoigne's reputation as the first inventor of the micrometer, if not as an aeronautical engineer, had been well established on both sides of the Channel. In his authoritative *Histoire de l'Astronomie Moderne* (1821)

[7] Birch, Thomas, *History of the Royal Society of London* (London, 1756), iii, H 482. 'Sir Jonas Moore related, that one Mr. Gascoigne had, above forty years before, made a contrivance for flying, by which he had been able to make a boy at Knareborough fly a considerable way; but that he being frightned in his flight by the acclamations of the spectators, fell down before he designed to alight, and though not much hurt, would not attempt it any farther'.

[8] *Correspondance du P.Marin Mersenne*, ed. Paul Tannery and Cornelis de Waard (Paris, 1932–1988), vol.IX see Mersenne to Pell, 20 Jan 1640, 53.

[9] *Ibid.*, vol.IX, Mersenne to Rivet, 15 July 1640 (trans. DS), 484.

[10] *Ibid.*, vol.IX, Descartes to Mersenne, 30 July 1640 (trans. DS), 524.

[11] *Ibid.*, vol.X, Descartes to Mersenne, 30 August 1640 (trans. DS), 87.

Table 6.1 Delambre's comparison of Gascoigne's solar radii

Date (o.s.)	Gascoigne	Connaissance des Temps
25 October	16′ 11″ or 10″	16′ 10″
31 October	16′ 11″	16′ 11″.4
2 December	16′ 24″	16′ 16″.8

the French astronomer, Jean-Baptiste Delambre, devoted several pages to Gascoigne and the priority dispute. He compared the Middleton astronomer's measurements of the radius of the solar disk favourably with the theoretical values given in the French almanac *Connaissance des Temps* (Table 6.1).[12]

Whilst giving due credit to Gascoigne, however, Delambre was dubious about the credence that could be given to the claims of his English protagonists—especially after the long hiatus between the receipt of Richard Towneley's letter and further follow-up by the Royal Society.

> These pieces of information suffice for those who would like to form their own judgement on the pretensions of those who can be regarded as the inventors of the micrometer. [They] …. prove one thing alone, that it was only on receipt of Auzout's letter that the Royal Society concerned itself with this matter. There was only one proof to be given, namely to produce, from the month of January [1667], the various instruments and the observations made in 1666 with the afore-mentioned instruments, showing the relationship between the different scales and minutes and seconds; in other words to do what Flamsteed or his heirs did 59 years later. We believe firmly in the reality of Gascoigne's observations, ….; but we believe neither in the observations of Towneley, who talks only of a scale of equal parts; nor in the observations of Hooke, who only had his instrument constructed several months after the discussion occasioned by Auzout's letter.[13]

The remark about Hooke chimed with Flamsteed's grumble, in relation to another dispute about optics, that Hooke 'has onely the conceite [i.e. the idea] not the instrument'.[14]

In 1830 Ralph Thoresby's diaries were published by the Rev. Joseph Hunter. Thoresby had been personally acquainted with relatives of the Gascoignes of Thorp-on-the-Hill—William's branch of the family. He was also a friend of Richard and Charles Towneley, with whom he had discussed the astronomer's life. In the diaries we learn for the first time the precise location of William's home. In an entry made during 1702 Thoresby describes a walk to

> Newhall, once the seat of the most celebrated mathematician, not only in these parts, but I believe in the world, viz. Mr William Gascoigne, eldest son of Henry Gascoigne, Esq. who so long ago as the time of King Charles the First, (in whose service he was slain,) discovered and made constant use of a curious instrument, that Monsieur Azout [sic], the French astronomer in this age, prides himself as the first inventor of.[15]

---

[12]Delambre, Jean-Baptiste, *Histoire de l'astronomie moderne* (Paris 1821), ii, 590.

[13]*Ibid.* (trans. DS), 592.

[14]Johns, Adrian, *Flamsteed's optics and the astronomical observer*, in *Flamsteed's stars*, ed. Frances Willmoth (Woodbridge, 1997), 84.

[15]Hunter, Joseph, *The Diary of Ralph Thoresby, FRS* (London, 1830), i, 357.

Fig. 6.2 Ralph Thoresby (From *The Diary of Ralph Thores by*, ed. J. Hunter, v.1, 1830)

He also gives an account of a visit to Towneley Hall the same year and conversations with the three brothers (Richard, Charles and John). He was shown much from their unique collection of manuscripts and instruments, but confesses to being 'best pleased with the collection of original letters that passed through Mr. Christopher Towneley, the antiquary's contrivance, between Mr. Gascoigne, of Yorkshire, and Mr. Crabtree and Mr. Horrox of Lancashire'.[16]

Only 2 years after the publication of the Thoresby Diaries, two volumes of his correspondence were published, under the title *Letters of eminent men addressed to Ralph Thoresby, FRS*. Included among the letters was one from Charles Towneley (16 January 1699), telling Thoresby (Fig. 6.2) that his brother Richard Towneley—by then almost blind, due to cataracts—had many of Gascoigne's papers. In it he also discloses that,

> My brother has been told by my uncle, that Mr Gascoigne, at his father's house when he was slain, had a whole barn full of machines or instruments; it is not known what he intended them for, but perhaps, if some of them could be found, guesses might be made which way his endeavours or further studies looked.[17]

Christopher, his uncle, the famed 'Transcriber', had seemingly visited New Hall after William's death to try to retrieve some of the astronomer's papers. Without his exertions, the remembrance of Gascoigne would surely have evaporated by this stage. Now, instead, by the early 1830s, Gascoigne's priority in respect of the invention of the telescopic sight and the micrometer had been settled internationally. Nevertheless, none of his letters had been published in its entirety—nor had any sustained explanation of his optical discoveries been brought to light. Almost 200 years after his death, however, this omission would be rectified by material prised out of the depths of a castle in Oxfordshire.

[16] *Ibid.*, 387.

[17] *Letters of eminent men addressed to Ralph Thoresby, FRS* (1832), i, 353.

Chapter 7
In His Own Hand

When the London doctor, John Bevis, reported to the Royal Society in 1753 that he had unearthed in the Macclesfield library at Shirburn an original letter from Gascoigne to William Oughtred, he could not have imagined that the letter would not resurface for almost another century. But that is indeed how things turned out. It was a pity, because, beyond a handful of astronomical measurements, Bevis quoted nothing from the nine-page long letter. Nor did he mention that this was the second letter to Oughtred from the Yorkshire astronomer and that the first letter too was in the library.

Unfortunately, although private collections serve to preserve many invaluable manuscripts, they tend to be inaccessible: They are not often open to casual browsers; nor are their contents routinely documented in published catalogues. In the early 1860s, the Macclesfield library was catalogued to a very high standard by Edward Edwards, but the catalogue itself remained simply as a manuscript in the library.[1]

Under the first two Earls of Macclesfield—Thomas Parker and his son George—the moated Shirburn Castle in Oxfordshire had become a centre of scientific renown. Besides the magnificent scientific library, they had also established here a chemical laboratory and a first-class astronomical observatory—perhaps the reason for the visit by Bevis. George, in particular, was a devotee of science, becoming President of the Royal Society from 1752 until his death.

Both the first and second Earls had a good grounding in mathematics. Their tutor was the Welsh mathematician William Jones (c.1675–1749), who lived at the castle for a considerable time. Jones had many claims to fame, including the fact that it was he who introduced the Greek letter π as the symbol for the ratio of the circumference of the circle to its diameter, but his role in the preservation of Gascoigne's story was crucial. He was an avid collector of books and letters. His collection of

[1] Quarrie, P., *The Scientific Library of the Earls of Macclesfield*, Notes and Records of the Royal Society (2006), v.20, 19.

D. Sellers, *In Search of William Gascoigne: Seventeenth Century Astronomer*, Astrophysics and Space Science Library 390, DOI 10.1007/978-1-4614-4097-0_7,

some 15,000 books was considered to be the most valuable mathematical library in England.[2] Upon his death in 1649, this library—which contained the Gascoigne letters—was bequeathed to George, the second Earl. There the letters remained, privately, for more than 350 years.

The seventh Earl, George Lovenden William Henry Parker (1888–1975), acted as guardian of the library for 79 years after he inherited the title in 1896. 'This posed a formidable problem to historians of science throughout most of the twentieth century', according to astronomy historian Owen Gingerich, 'because the 7th Lord Macclesfield had fended off historians. It was only by a special concession that the Royal Society, Britain's leading scientific organization, had been allowed to copy the rich repository of Isaac Newton letters in the collection in order to complete its magisterial seven-volume edition of Newton's correspondence.'[3]

The Macclesfield library was ultimately dispersed in 2004–2005, as the result of financial problems and an acrimonious family dispute that saw the ninth Earl evicted from Shirburn Castle. Up to that time, however, almost the only permitted publication of scientific manuscript material from the library, apart from the Newton letters, was the substantial body of transcriptions made by the Oxford astronomer, Stephen Peter Rigaud (1774–1839). Rigaud died before publication, but the 1,000 or so pages were seen through the press by his son Stephen Jordan Rigaud, later to become bishop of Antigua. They appeared in 1841 as a two-volume book entitled *Correspondence of scientific men of the seventeenth century*. The first volume included the two letters from William Gascoigne to the mathematician, William Oughtred.

William Oughtred (c.1575–1660) was not a professional mathematician: He was a clergyman—the rector of Albury in Surrey—and most of the mathematical instruction that he delivered to young enthusiasts was done without thought of charging.[4] His considerable reputation was built on his 1631 book, *Arithmeticae in numeris et speciebus institution: quae tum logisticae, tum analyticae, at que adeo totius mathematicae quasi clavis est* ('The method of calculating in numbers and letters: which was the key first to arithmetic, then to analysis and now to the whole of mathematics'[5])—or *Clavis mathematicae* (Mathematical Key) for short. This tiny book, only 88 pages long and scarcely 11 cm wide, was a concise guide to the latest algebraic methods. The first edition, which had been studied avidly by Gascoigne, was in Latin, but in 1647 and later English editions appeared and it remained a classic amongst students of mathematics for most of the seventeenth century. As a result, it is no surprise that Oughtred was the recipient of letters from earnest strangers seeking his opinion or help. The letter of June 1642 that William Price sent from London to Oughtred gives a flavour of the demands brought by fame:

---

[2] Wallis, R., *Jones, William (c.1675–1749)*, Oxford Dictionary of National Biography (2004), online edition.

[3] Gingerich, O., *The Book Nobody Read* (New York, 2004), 91–92.

[4] Cajori, F., *William Oughtred* (Chicago, 1916), 7.

[5] Stedall, J.A., *Ariadne's Thread: The Life and Times of Oughtred's* Clavis, Annals of Science, v.57 (2000), 29.

Fig. 7.1 William Oughtred (Courtesy of the Thomas Fisher Rare Book Library, University of Toronto)

> Sir, Though I am a stranger to your person, yet I am well acquainted with the fame of your singular skill in the mathematics, and thereupon have so far presumed to intreat your assistance for the geometrical solution of the inclosed diagram ....[6]

Oughtred was by then almost 70 years old and, becoming aware of increasing frailty (though he was to live into his mid nineties), was reluctant to venture into new mathematical projects (Fig. 7.1). Besides, he felt that preferment in his career had been denied to him because his ecclesiastical peers took a dim view of the time he had spent on mathematics. Perhaps he ought to have curtailed his learned pursuits, but his love of mathematics was too great. Intrigued by Price's problem, he supplied a detailed response 4 days later, which began:

> Sir, It is true that I have bestowed such vacant time, as I could gain from the study of divinity, (which is my calling,) upon human knowledges, and, amongst other, upon the mathematics, wherein the little skill I have attained, being compared with others of my profession, who for the most part contenting themselves only with their own way, refuse to tread these salebrous and uneasy paths, may peradventure seem the more. Now being in years and mindful of mine end, and having paid dearly for my former delights both in my health and state, besides the prejudice of such, who not considering what incessant labour may produce, reckon so much wanting unto me in my proper calling, as they think I have acquired in other sciences; by which opinion (not of the vulgar only) I have suffered both disrespect, and also hinderance in some small perferments I have aimed at. I have therefore now learned to spare myself, and am not willing to descend again *in arenam*, and to serve such ungrateful muses. Yet, sir, at your request I have perused your problem ...[7]

By any reckoning Oughtred worked hard. 'His head was always working', declared the diarist John Aubrey, 'He would drawe lines and diagrams on the dust ... Studyed late at night; went not to bed till eleaven a clock; had his tinder box by him; and on the top of his Bed-staffe, he had his Inke horne fix't. He slept but little. Sometimes he went not to bed in two or three nights, and would not come downe to meales till he had found out the *quaesitum*.' But, confirming what the old man thought others made of him, Aubrey added, 'I have heard his neighbour Ministers say that he was a pittiful Preacher; the reason was because he never studied it, but bent all his thoughts on the Mathematiques'. If Aubrey is to be believed, Oughtred's wife 'was a penurious woman, and would not allow him to burne a candle after Supper, by which meanes many a good notion is lost, and many a Probleme unsolved'.[8]

Gascoigne's first letter to Oughtred (Fig. 7.2), dated 2 December 1640, opened in the same adulatory vein as that of Price:

> Sir, Amongst the mathematical rarities these times have afforded, there are none of that small number I (a late intruder into these studies) have yet viewed, which so fully demonstrates their authors' great abilities as your Clavis ...[9]

---

[6] Rigaud,S.J., *Correspondence of scientific men of the seventeenth century* (Oxford, 1841), i, 59.

[7] *Ibid.*, 60–1.

[8] Dick, O.L., *Aubrey's Brief Lives* (New Hampshire, 1999), 222–5.

[9] Rigaud, S.J., *Op.Cit.*, 33.

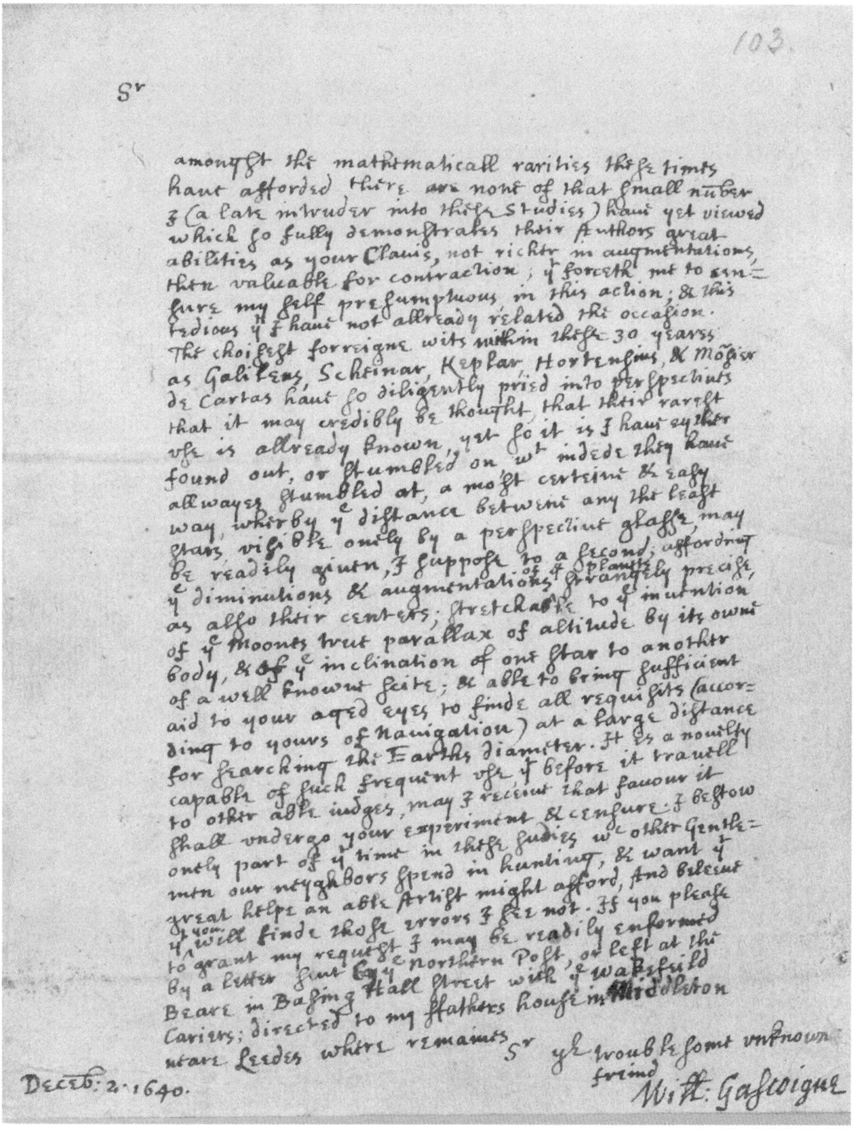
103

Sr

amongst the mathematicall rarities these times
haue afforded there are none of that small nūber
I (a late intruder into these studies) haue yet viewed
which so fully demonstrates their Authors great
abilities as your Clauis, not richer in augmentations,
then valuable for contraction; ye forceth me to cen=
sure my selfe presumptuous in this action; & this
tedious if I haue not allready related the occasion.
The choisest forreigne wits within these 30 yeares
as Galileus, Scheinar, Keplar, Hortensius, & Mosier
de Cartas haue so diligently pried into perspectiues
that it may credibly be thought, that their rarest
use is allready knowne, yet so it is I haue eyther
found out, or stumbled on wch indede they haue
allwayes stumbled at, a most certeine & easy
way, wherby ye distance betweene any the least
stars visible only by a perspectiue glasse, may
be readily giuen, I suppose to a second; affording
ye diminutions & augmentations of ye planets strangely precise,
as also their centers; stretchable to ye inuention
of ye moones true parallax of altitude by its owne
body, & of ye inclination of one star to another
of a well knowne seite; & able to bring sufficient
aid to your aged eyes to finde all requisits (accor=
ding to yours of nauigation) at a large distance
for searching the Earths diameter. It is a nouelty
capable of such frequent use, yt before it trauell
to other able iudges, may I receiue that fauour it
shall undergo your experiment & censure. I bestow
onely part of ye time in these studies wch other Gentle=
men our neyghbors spend in hunting, & want ye
great helpe an able Artist might afford, And beleeue
ye you will finde those errors I see not. If you please
to grant my request I may be readily enformed
by a letter sent by ye northerne Post, or left at the
Beare in Basing Hall street with ye Wakefeild
Cariers; directed to my fathers house in Middleton
neare Leedes where remaines Sr yr troublesome unknowne
freind
Will: Gascoigne
Decemb: 2. 1640.

Fig. 7.2 Gascoigne's first letter to Oughtred (Reproduced by kind permission of the Syndics of Cambridge University Library (Add.9597/13/5/103,r))

He then simply enquired whether Oughtred would be willing to review a description of a means

> Whereby the distance between any the least stars, visible only by a perspective glass, may be readily given, I suppose to a second; affording the diminutions and augmentations of the planets strangely precise, as also their centres; stretchable to the invention of the moon's

true parallax of altitude by its own body, and of the inclination of one star to another of a well known site.[10]

Oughtred's generosity was well-known and the young astronomer felt that a favourable opinion from the aging mathematician would give him the confidence that he needed to approach other leading scientists.

It is a novelty, capable of such frequent use, that before it travels to other able judges, may I receive that favour it shall undergo your experiment and censure?[11]

Gascoigne must have received an encouraging response from the old man, for the undated second letter (written c.February 1641)—the same one that Bevis had spoken of: nine pages long in its manuscript form—gives an extended description of the young astronomer's experiments in determining the rules of refraction, the means of calculating a ray's progress through optical devices, his invention of the telescopic sight and the micrometer, together with a number of observations. To illustrate the explanations he inserted a page of five diagrams that would also accompany several of his future letters to others.

Here he describes his experiments to verify the sine law of refraction and uses this to determine the path of light rays through plano-convex and plano-concave lenses. The properties of the lenses in a galilean telescope he uses to calculate the quantitive relationship between the size of a projected image of the sun and the angle subtended by the solar disk in the sky. Example measurements are given of the semi-diameters at perihelion and aphelion (16′ 25″ and 15′ 53″ respectively), which convince him of the bisection of the eccentricity of the Earth-Sun orbit—thus lending support to Kepler against Lansberge and Hortensius. Most remarkably, he reveals the serendipitous discovery that two objects—a distant one, being viewed through the telescope, and a proximate one, within the telescope—could both be in sharp focus at the same time:

This is that admirable secret, which, as all other things, appeared when it pleased the All Disposer, at whose direction a spider's line drawn in an opened case could first give me by its perfect apparition, when I was with two convexes trying experiments about the sun, the unexpected knowledge.[12]

As Gascoigne pointed out, two convex lenses were essential to this result. This was the format of telescope proposed by Kepler, wherein the objective lens was convex, but, unlike in the Galilean model, so too was the eyepiece lens. With the Keplerian arrangement he found that any object placed at the focal point of the objective lens, could be magnified and brought into focus alongside the image of the distant object being viewed. This 'incredible rarity', as he styled it, immediately

[10] *Ibid.*, 33. The original Gascoigne-Oughtred letters were acquired by the University of Cambridge Library in 2000. *Macclesfield MSS*: 2 December 1640, Add.9597/13/5/103; c.February 1641, Add. 9597/13/5/104-9.

[11] *Ibid.*, 33–34.

[12] *Ibid.*, 46.

suggested a number of possibilities. Firstly, it was feasible to introduce a simple marker, to act as a telescopic sight.

> if I .... placed a thread where that glass [the eyepiece] would best discern it, and then joining both glasses, and fitting their distance for any object, I should see this at any part that I did direct it to ....[13]

Secondly, one could introduce a measuring device into the common focal point and thereby measure the size of the observed image.

> Upon trial I found ... that the moon's picture here was in every respect answerable, for apparition of its diameter, to an object of equal rad. placed here [at the common focal point]. Thence I begun to contrive for some excellent scale, which might afford me in as small parts, as were possible, *ql* [a distance in the image plane[14]], believing that art could not afford such distinct divisions as might be severally distinguishable in *ql*. At last I contrived this many barred scale[15]

Unfortunately, no drawing of Gascoigne's scale was found with the letter to Oughtred. Nevertheless, it is clear that he used the scale, albeit 'imperfect', to measure lunar diameters and infer the eccentricity of the lunar orbit. Furthermore, he had fitted 'glasses and a hair'—a telescopic sight—to his quadrant.

Finally, in this whistle-stop tour of his discoveries, Gascoigne describes how he has lately replaced the scale with a device incorporating moveable pointers, mounted on screws: a micrometer. With this, in the autumn and winter of 1640, he has measured the diameters of Jupiter, Mars and Saturn to an impressive degree of accuracy, but stresses the limits imposed by having to put up with 'such ill-suited glasses as I yet am worth'.

Towards the close of his letter he tells Oughtred how he has begun to divulge details of his optical inventions to others:

> I have shewed some part of these to two of my very late acquaintance, both industrious astronomers, who are not a little taken with them, and the more, because they believed they had tried all possible means, that glasses could afford for the measuring of diameters, yet never attained to any more than guessing, and indeed were so confident, as they would not believe until I let them see a glass to try the moon's diameter by, and how I proceeded in the other usual one. Indeed if one of my noble friends had not, before I intended it, bolted out so much as it could no longer lie hid, I had reserved it until I had got some rare glasses worthy of a prince's inspection, unto whose notice, at his majesty's being at York, he had, unknown to me, promoted this invention, who, as I conceive his words, will expect an ample relation. He also acquainted me with the famous Sir Kenelm Digby, unto whom I first shewed the contrivance of the scale, and its use in a glass, whose courtesy bound me to fulfil his request for some of the diagrams and observations, which I have lately sent him.[16]

The 'industrious astronomers' in question were William Crabtree and a friend (most probably Christopher Towneley), who had visited him at Middleton sometime prior to the end of October 1640.

---

[13] *Ibid.*, 46.

[14] The diagrams accompanying this letter are included in Chap. 18 (see Diagram 5).

[15] *Ibid.*, 46–7.

[16] *Ibid.*, 54.

In the same year that the Gascoigne-Oughtred letters were published by Rigaud, one of his young acquaintances, James Orchard Halliwell (1820–1889), brought out another collection of scientific correspondence. *A collection of letters illustrative of the progress of science in England* was only 1 of 13 books prepared for publication that year by the prolific 21 year old. Its principal interest in relation to Gascoigne lay in the inclusion of a letter of 26 August 1644 from Sir Charles Cavendish (1591–1654)—freshly exiled in Hamburg following the battle of Marston Moor—to fellow mathematician John Pell, in which he wrote,

> I am well acquainted with Mr Gascoine, whoe was providore to oure armie: he is an ingenious man & hath shewed me howe perspectives[17] maye be much improved; I onelie mislike his glass next the eye which he makes convex on both sides; I tolde him it woulde make confused sight, if De Cartes his doctrine be true, but upon triall it proved more distinct than I expected, yet I thinke a concave on that side next the eye would doe better; his perspective did not multiplie more than myne as I thinke, but his speculation is most true, and this was one of his first trialls and not made to the manner of his best inventions.[18]

Thus, from a relatively trustworthy source we learn of Gascoigne's role in the royalist army. We also see evidence of a broadening circle of Gascoigne's acquaintances. Cavendish was in touch with many of the most capable mathematicians in England, as well as continental natural philosophers, such as Marin Mersenne, René Descartes and Claude Mydorge.[19]

Maybe Gascoigne's acquaintance with Charles Cavendish also provides a clue as to the identity of the garrulous "noble friend" mentioned in the letter to Oughtred. It is tempting to speculate that this was the brother of Charles: William Cavendish, Marquess, and ultimately Duke, of Newcastle. In several ways Newcastle would be the likeliest candidate. In May 1638 he had been appointed as governor of the 8-year-old Prince Charles. The prince's father, Charles I, left London for York on 20 August 1640 and on 24 September convened there a Great Council of Peers, which

---

[17] *Perspectives*: i.e. telescopes.

[18] Halliwell, James Orchard, *A collection of letters illustrative of the progress of science in England* (London, 1841), *Sir Charles Cavendish to John Pell*, Aug 16/26 1644, 83. Also, Charles Cavendish to John Pell, 16/26 August 1644, British Library MS Add. 4278, ff.180–1 (transcribed in *John Pell (1611–1685) and his correspondence with Sir Charles Cavendish* by Noel Malcolm and Jacqueline Stedall (Oxford, 2005), Letter 14, 369–70). Note: the same letter transcribed in *Correspondance du Père Marin Mersenne*, ed. by Cornelis de Waard *et al.* (Paris, 1932–88), xiii, 204–5, is incomplete and omits the piece about Gascoigne. One wonders whether the antecedent letter from Pell to Cavendish in Malcolm and Stedall's work is also incomplete, since Pell does not ask any question about Gascoigne, even though Cavendish appears to be answering one. Although the letter was written a month after the Battle of Marston Moor, in which Gascoigne was probably killed, Cavendish does not seem to be aware of Gascoigne's demise.

[19] Jacquot, Jean, *Sir Charles Cavendish and his learned friends*, Annals of Science (London, 1952), v.8, 13–27, 175–191.

Newcastle probably attended.[20] It is a matter of record that both William and Charles Cavendish were passionately interested in telescopes and were enthusiastic purchasers of lenses. Indeed, William was presented with a large Italian telescope by Kenelm Digby in 1648.[21] His interest in astronomy too was longstanding. As early as 1633 he was urging the philosopher, Thomas Hobbes, to obtain for him, within a year of its first publication, a copy of Galileo's *Dialogue on the two great world systems* (1632).[22] What is more, William Cavendish was to acquire a reputation for 'bolting things out': in 1658, during the time of the Protectorate, when leading royalists were in enforced exile, he carelessly divulged that one of their number, the Earl of Ormonde, was undertaking a dangerous secret journey into England—other exiles were aghast 'that anyone who knows Newcastle would trust him with so important a secret, for it might as well be proclaimed at the cross'.[23]

After the 1841 publications of Rigaud and Halliwell, little new material relating to Gascoigne was to appear.

From time to time local historians in Yorkshire raked over the accounts of Gascoigne's lineage and his role in the civil war, but discovered virtually nothing new. The most assiduous of the researchers was William Wheater, who in 1863 published a short essay in *The Gentleman's Magazine*. The main thrust of his essay[24] was an appeal for help in digging up material that might help him in his 'attempt to rescue the memory of this long-neglected genius from the undeserved oblivion into which it appears to have fallen'. Biographical errors and inconsistencies of previous accounts, including Derham's, were brought under the spotlight. Amendments here and there were culled from the Gascoigne pedigree that had been drafted, but never brought to print, by the local antiquary—and contemporary of Gascoigne—John Hopkinson.

[20] House of Lords, file ref.: HL/PO/JO/10/1/42, 24 and 25 September 1640 – List of peers present at the Council of York on the 24th. This list of 62 names does not include Newcastle, but a footnote says "some other Lords have come in since, but this was the Counsell of the first day".

[21] Charles Cavendish to John Pell, 4/14 February 1648, British Library MS Add. 4278, ff. 271–2 (transcribed by Noel Malcolm and Jacqueline Stedall, *Op. Cit.*, Letter 72, 502–3).

[22] Historical Manuscripts Commission, 13th Report, Appendix, Part II (HMSO, London, 1893): *The manuscripts of His Grace the Duke of Portland, preserved at Welbeck Abbey*, ii, 124, Letter from Thomas Hobbes to the Earl of Newcastle, 26 January 1633. Hobbes wrote: My first businesse in London, was to seeke for Galileo's Dialogues; I thought it a very good bargain, when at taking my leave of your Lordship I undertooke to buy it for you, but if your Lordship should bind me to performance it would be bad enough, for it is not possible to get it for money. There were but few brought over at first, and they that buy such bookes, are not such men as to part with them againe. I heare say it is called in, in Italy, as a booke that will do more hurt to their religion then all the bookes have done of Luther and Calvin, such opposition they thinke is between their religion, and naturall reason.

[23] Calendar of State Papers, Domestic Series, 1657–1658, ed. Green,M.A.E. (London, 1884), 300. Statement of Thomas Ross to Secretary Nicholas (24 February 1658, OS).

[24] Wheater, William, *Gentlemans magazine*, ccxiv (1863), 760–2.

Virtually every subsequent account of Gascoigne seems to have been confined to an examination of the material published before this time, or to citing secondary sources that were themselves based on that material. Strangely, the substantial correspondence that was exchanged by Gascoigne and Crabtree—though transcribed at length by Flamsteed—remained unpublished, either whole, or in part: Only the portions extracted by Derham and the two letters reproduced by Rigaud appear generally to have been quoted.[25]

[25] Eric G.Forbes, however, does quote an extensive abstract from Crabtree's letter of 21 June 1642 from the Flamsteed papers in *The Gresham lectures of John Flamsteed* (London, 1975), 51–2.

Part II
Gascoigne's World

Chapter 8
The Religious World of William Gascoigne

Delving further into the life and work of William Gascoigne, checking the reliability of the mixture of hagiography and supposition that surrounds him, enquirers must try to enter his world: To enter one of the most turbulent periods in the religious history of England—a time of fear, persecution, mistrust and secrecy.

Recusancy

It is commonly claimed that Gascoigne was a Roman Catholic. Almost all recent secondary accounts are emphatic about this[1] and make much of the fact that this did not hinder his friendly relationships with William Crabtree and Jeremiah Horrocks, who are deemed to be of a different persuasion. Maybe this is a fair assumption, given John Aubrey's claim that Gascoigne was 'bred up by the Jesuites'.[2] After all, several other branches of the Gascoigne family in the region—the Gascoignes of Barnbow, Gawthorpe and Lasingcroft—were undoubtedly prominent adherents of the 'old religion'. Nevertheless, hard facts are elusive. Their obscurity might have been intended.

If the Gascoignes of Thorp-on-the-Hill were Catholics, they would have certainly felt 'under seige' in post-Reformation England. During William's lifetime, most of the practices of the Roman Catholic religion were deemed to be crimes. The law forbade attendance at Mass, or the possession of a crucifix, rosary or prayer book.

Catholics and others refusing to attend Church of England services at the local parish church were labelled 'recusants'. The crime of 'recusancy' was spawned by

---

[1]E.g. Chapman, A., *Oxford Dictionary of National Biography (Vol. 21)*, (Oxford, 2004), 502: 'Almost certainly, Gascoigne belonged to the Roman Catholic gentry of Yorkshire.'

[2]Aubrey, John, *Brief Lives and other selected writings*, ed. Anthony Powell (London, 1949), 148.

D. Sellers, *In Search of William Gascoigne: Seventeenth Century Astronomer*, Astrophysics and Space Science Library 390, DOI 10.1007/978-1-4614-4097-0_8,

the *Act for the uniformity and administration of sacraments throughout the realm* of 1549 (2 and 3 Edward VI, c.1). This law introduced a Communion Service fundamentally at variance with the traditional Sacrifice of the Mass and forbade the use of 'any other rite'. Imprisonment was prescribed for anyone officiating at a different form of public service. Fines and imprisonment could be visited on anyone 'procuring' such a service. There was no penalty under the Act, however, for simply refusing to attend the new service.

In 1552, Parliament passed a second *Act of Uniformity* (5 and 6 Edward VI, c.1) imposing a revised *Book of Common Prayer* and, for the first time, making it compulsory to be present at the new service. Refusal to attend could result in punishment by 'the Censures of the Church'—maybe excommunication, or refusal of the right to be buried in the church grounds. Anyone convicted of 'hearing or being present at' any other form of service, however, could be imprisoned. For a third such conviction, the penalty was imprisonment for life.

These Acts were repealed during the 5 year reign of Mary (1553–1558), herself a Catholic, but in 1559 under her sister, Elizabeth I, a further *Act of Uniformity* (1 Elizabeth, c.2) was introduced, in which fines were added in place of, or in addition to, the custodial penalties of the earlier Acts.[3]

For landed Catholics, such as the Gascoignes might have been, the stakes were high. Inability to meet the fines could lead to forfeitures of property and seizure of land. Lack of property or land to cover the fines would lead to imprisonment.

The harsh treatment meted out to John Towneley of Lancashire was an example of what recusants could expect. Under a 1601 portrait of Towneley, a prominent Catholic, whose grandsons were to be instrumental in preserving what little we know of Gascoigne, is the inscription:

> For professing the Apostolical Roman Catholic Faith, he was imprisoned first at Chester Castle, then sent to the Marshalsea, then to York Castle, then to the Blockhouses in Hull, then to the Gatehouse in Westminster, then to Broughton in Oxfordshire, then twice to Ely in Cambridgeshire; and so now, seventy-three years old and blind, is bound to appear and to keep within five miles of Towneley his house; who hath since the statute of 23rd Elizabeth paid into the Exchequer £20 a month, and doth still, that there is paid already about £5000.[4]

The stakes became higher and higher. Recusants risked not only their property and liberty, but eventually life itself. In 1563 the death penalty was imposed on any priest saying Mass. Under an Act of 1581—an *Act to retain the Queen's Majesty's subjects in their due Obedience*—the hiding of a priest was considered to be an act of Treason—a capital offence, punishable by death. Indeed, Margaret Clitherow, of the Shambles in York, was martyred, crushed to death between stones in 1586, for refusing to plead when on trial for concealing priests.

---

[3] Bowler, Dom Hugh, *Recusant Roll No.2 (1593–1594)*, Catholic Record Society, 1965, x–xi.

[4] Reynolds, E.E., *The Roman Catholic Church in England and Wales: A Short History*, (Wheathampstead, 1973), 242–243.

Equivocation

Little wonder that many Catholics, when interrogated about their faith, would adopt a policy of *equivocation*: giving the outward appearance of assenting to the doctrine of the Church of England, whilst inwardly remaining true to their faith. Invariably they would be 'economical with the truth'—taking care to make neither admissions nor denials. Detailed instructions about how, faced with inquisitors, 'without a lie a truth may be covered' were carefully described in the little *Treatise of Equivocation* published by the Jesuit priest, Henry Garnet. Thus, for example, 'we may use some equivocall word which hath many significations, and we understand it in one sense, which is true, although the hearer conceive the other, which is false'.[5] So, if asked whether a priest was hiding in a particular house, it would be legitimate for a Catholic to answer: 'You won't find him there!' The interrogator, it was hoped, would take the answer to be a plain denial, whereas it was an optimistic prediction rather than an answer.

Between 1581 and 1603 60 recusants and 120 priests were executed. After the Gunpowder Plot in 1605, anti-Catholic sentiments strengthened. Penal laws were reinforced and vigorously enforced. The Plot had been a reckless conspiracy by a number of Catholics to blow up the Houses of Parliament, along with the King and many Members of Parliament. The plot was discovered before it could be implemented and all the conspirators were gruesomely tortured and executed. Guy Fawkes, now the best known of the conspirators, was baptised in the same church—St Michael-le-Belfrey, in York—where, 5 years before the plot, Gascoigne's parents had married. Father Henry Garnet, found guilty by association, since he had learned of the plot through the confessional box, was hanged in front of St. Paul's Cathedral on 3rd May 1606. His trial and the condemnation of 'equivocation' that attended it had a deep impact on the popular attitude in England towards Catholics and towards Jesuits in particular. The matter even entered the theatre. Shakespeare wrote his play *Macbeth* in the very year that Garnet was hanged and has the Porter, supposing himself welcoming someone at the approach to hell-gate with the words,

> Faith, here's an equivocator, that could swear both the scales against either scale, who committed treason enough for God's sake, yet could not equivocate to heaven: O, come in equivocator.[6]

Although the persecution of Catholics in England diminished substantially during the eighteenth century, their formal emancipation did not come until 1829.

After the passage of the 1581 Act, failure to attend Church of England services became an indictable offence. The fines levied were recorded in a series of rolls, called the Recusant Rolls. These can still be examined in the National Archives, but as yet cast no further light on the religious inclination of William and his immediate family. The Rolls covering the period between 1592 and 1596 have been transcribed and published by the Catholic Record Society. A distant relative, also named William Gascoigne of

---

[5]Jardine, David (ed.), *A Treatise of Equivocation* (London, 1851), 48.

[6]Shakespeare, William, *Macbeth* (London, 1807), Act2, Scene 1.

Thorpe super Montem [Thorp-on-the-Hill], has several entries to his credit and in the 1593–1594 roll is recorded as being fined £140.[7] Despite the hefty fine, this William remained defiant and is mentioned, along with his servants, in a 1604 list of recusants and noncommunicants now preserved in the Bodleian Library[8]:

> William Gascoigne, Ellynor his supposed wife, Bridget Hunt, Mary Hemsworth: Recusants for 12 yeares.
>
> John Chamberlane in house with ye said Gascoigne; William Hargrave son of William Hargrave; Recusants 2 years.
>
> Jayne Leigh serunt to ye said William Gascoigne, George Parker; noncommunicants.
>
> *Secretly married* William Gascoigne aforesaid, Elynor his wife.

Records from 1595 report that 'Wm Gascoigne Junior coom not diligently to church'. A William Gascoigne was again cited as a recusant in 1615.[9]

Unfortunately, the Recusant Rolls of the West Riding of Yorkshire Quarter Sessions for most of the seventeenth century have been lost due to dampness. In any case, there is a further problem with looking to the Recusant Rolls for an answer to the question of whether our William was a Catholic.

Being labelled as a recusant would bring crippling financial penalties for the Catholic Gentry. In order to avoid these, heads of families and their eldest sons would often comply with the legal requirement to attend church services. They would become 'church papists', while their wives and younger children faced the wrath of the state as recusants. Compromises of this sort make it difficult to find conclusive evidence that anyone was, or was not, an adherent of the 'old religion'. The burial of William's mother, Jane, at Rothwell Parish Church is not proof that the family was protestant, since the Catholic gentry often had their deceased buried according to the Protestant rights. The fact that his parents, Jane and Henry, were married in a parish church would be more unusual, if they were Catholics,[10] but still could be explained as an attempt to ensure that financial arrangements and inheritance rights were not jeopardised by an unlawful marriage. Henry also married his second wife, Grace, at Rothwell Parish Church.

On 18 July 1639 at the West Riding Sessions in Wakefield a case was brought by four Rothwell parishioners complaining that the churchwardens had failed to dole out the poor relief that was due to them. The court received as counter-evidence a certificate made out by the churchwardens and overseers, together with Sir Ferdinando Leigh, the vicar, Edmund Kay, and Henry Gascoigne, Esquire.[11] The Gascoigne's were amongst the well-to-do of the parish and seemingly did not shirk involvement in its affairs.

---

[7] Bowler, Dom Hugh, *Op.Cit.*, 218.

[8] Peacock, Edward, *A List of the Roman Catholics in the County of York in 1604*, (London, 1872), 13.

[9] Aveling, Dom Hugh, *The Catholic Recusants of the West Riding of Yorkshire: 1558–1790*, Proceedings of the Leeds Philosophical and Literary Society, Literary and Historical Section, v.10, pt.vi (Leeds, 1963), 301.

[10] Cliffe, J.T., *The Yorkshire Gentry from the Reformation to the Civil War* (London, 1969), 204.

[11] The West Riding Session Rolls, *The Yorkshire Archaeological and Topographical Journal*, v.5 (1879), 394.

Education

The character of the schooling received by children in seventeenth century England would often reveal the religious affiliation of their families, but little is known of William Gascoigne's education. In the second letter to Oughtred, Gascoigne talking of his astronomical knowledge, says

> I entered upon these studies accidentally after I betook myself to the country, having never had so much aid as to be taught addition, nor the discourse of an artist (having left both Oxford and London before I knew what any proposition in geometry meant) to inform me what were the best authors; nor being rich enough in language to understand perfectly any tongue except my mother's, having had two year's interim between the school and university, where I dare not say that I learned more than how those lived, that increased their knowledge, and to know my own wants by others' wealth, and to be sensible that the judgement at Babel was no trifle.[12]

In common with many male offspring of the gentry in the early seventeenth century, William Gascoigne is likely to have received a thorough training in classical learning at school. But where would his education have been received? Many members of the Catholic gentry sent their sons to Catholic colleges abroad. Does Aubrey's statement about being 'bred up by the Jesuites' suggest that Gascoigne had gone down this route? Certainly, a William 'Gascoigne' or 'Gaskin' did enter St.Omers College, run by English Jesuits near Liège, in 1629 or earlier.[13] Our William would have been 17 years old: A not inappropriate age for such a step. John and Charles Towneley, for example, entered Douai College (France) in 1649, at the ages of 18 and 19 respectively.[14] But Gascoigne's confession that he could not 'understand perfectly any tongue except my mother's', does not seem consistent with a continental education.

Gascoigne evidently felt sorely hampered by the 'judgement at Babel' where, according to Genesis (ch.11,v.1–9), the Lord 'did confound the language of all the earth' and thereby thwarted the overweening ambition of his people by rendering them unable to understand one another's speech. We need to make allowance, however, for Gascoigne's deep, but unwarranted, modesty about his own abilities. His few remaining letters and manuscripts make it clear that he could read and translate Latin texts by Kepler and Lansberg. He was also familiar with current works of French mathematicians, which by that point had not been published in English. Despite his protestations, therefore, it seems he was not confined to his mother's tongue.

At the time that Gascoigne's family might have been considering a continental education for their eldest son, the law required anyone going abroad to have a pass

---

[12]Rigaud, S.P., *Correspondence of Scientific Men of the Seventeenth Century* (Oxford, 1841), v.1, 35.

[13]Holt, Geoffrey, *St.Omers & Bruges Colleges, 1593–1773: A Biographical Dictionary* (London, 1979), 111.

[14]Webster, C., *Richard Towneley (1629–1707), the Towneley Group and Seventeenth Century Science*, Transactions of the Historic Society of Lancashire and Cheshire, v.118 (1967), 62–63.

issued by the Privy Council. In 1625 the Privy Council directed that every such pass should stipulate that no visit to Rome was permitted. Between the years 1619 and 1631, however, the Privy Council issued no such pass to a 'William Gascoigne'. If he did acquire his education overseas, his passage from England, like most others, must have been in conditions of clandestinity, or under an alias.

For Catholics, an alternative to education in France or the Low Countries was sometimes provided by Rome. John Aubrey, in his biographical sketch of Jonas Moore, asserts that Gascoigne was 'bred by the Jesuites at Rome'. The Venerable English College in Rome, however, does not record in its annals (the *Liber Ruber*) that anyone matching the description of Gascoigne was registered as one of its students during the first half of the seventeenth century.[15] It is tempting to speculate that the Roman College, set up by Ignatius Loyola—the founder of the Jesuit order—was instead the scene of William's education. After all, Jonas Moore's story, passed on by Aubrey, seems to suggestion a connection: 'Sir Jonas told us that a Jesuite (I think it was Grenbergerus, of the Roman College) found out a way of flying, and that he made a youth performe it. Mr. Gascoigne taught an Irish boy the way …'. Christoph Grienberger (1561–1636) succeeded Christopher Clavius as a professor of Mathematics at the Roman College. He was an astronomer of note, the inventor of the equatorial telescopic mount, an investigator of many species of mechanical device, and possibly a Copernican at heart.[16] Yet it is scarcely credible, had he been taught by Grienberger, that Gascoigne would have had to rely on his own resources to learn geometry, mathematics and astronomy.

The fact that Catholics were excluded from the English Universities—at the time only Oxford and Cambridge—was one of the main reasons for them choosing to study abroad. It is unlikely, however, that William Gascoigne would have suffered such exclusion. Two years after the death of his first wife, Jane, William's father, Henry, decided to become a mature student at Oxford. He matriculated at Queen's College on 7th May 1619 at the age of 32,[17] though there is no indication that he completed his degree. William was still a small child—scarcely 7 years old. Presumably, Henry lived in Oxford, perhaps with his young family, whilst engaged in his studies.

By the late 1620s Gascoigne's family were back in the parish of Rothwell—if indeed they had ever been away. A fresh start was inaugurated by the remarriage, in the summer of 1629, of his father, Henry. His new wife, Grace, one might suppose was not much older than the 17 year old William, to whom she now became

[15] Kelly, Wilfrid (Ed.), *Liber Ruber Venerabilis Collegii Anglorum de Urbe* (v.1, 1579–1630, & v.2, 1631–1783), Catholic Record Society (London, 1940).

[16] Lattis, James M., *Between Copernicus and Galileo: Christopher Clavius and the Collapse of Ptolemaic Cosmology* (Chicago, 1994), 203–5.

[17] Foster, Joseph, *Alumni Oxoniensis: the members of the University of Oxford, 1500–1714* (London, 1891), 552. The entry reads "*Gascoigne, Henry, of co. York, arm. QUEEN'S COLL., matric. 7 May, 1619, aged 32; of Thorpe-on-the-Hill, Yorks (s. John), clerk of the peace for the West Riding; baptised 19 Nov., 1586; buried 20 Nov., 1645; father of John 1634.*"

stepmother. Six children arrived in quick succession between 1631 and 1639: Thomas, Maria, Richard, George, Margaret and Ellinor—half-brothers and half-sisters for William. Sadly, and not unusually for the times, four of these failed to reach their 15th birthday. That the name Margaret was given to Grace's fifth child indicates, perhaps, that William's older sister of the same name had died before completing her teenage years.

Sometime after the finishing of his formal education, wherever that might have been, William Gascoigne too returned 'to the country', presumably to his father's household in Middleton.[18] At roughly the same time his youngest full brother, John, by then 17 years old, went off to study at University College, Oxford, where on 23rd April 1638 he gained a BA degree.[19] It is clear that there was no practical impediment that debarred the Gascoignes from higher education in England, even though William himself does not feature in the matriculation registers of Oxford or Cambridge. Given his claim to have attended a University, it must have been informally at one of these establishments—most likely Oxford.

Protestation

As we have seen, many Catholics preferred to 'equivocate' in matters regarding religious affiliation. This may have been the approach of our William and his father when they were asked—along with other citizens—to take the *Protestation Oath* in 1641. The Oath included the following text:

> I, [*name*], do in the presence of Almighty God, promise, vow and protest, to maintain and defend, as far as lawfully I may, with my life, power, and estate, the true reformed Protestant religion, expressed in the doctrine of the Church of England, against all popery and popish innovations within this realm, contrary to the same doctrine, and according to the duty of my allegiance, his majesty's royal person, honour and estate; as also the power and privileges of Parliament, the lawful rights and liberties of the subject, and every person that maketh this Protestation, in whatsoever he shall do in the lawful pursuance of the same…

In the House of Lords record of 'The names of all such within the townshipp of Middleton in the parish of Rothwell as have taken the protestation' the names of Henry Gascoigne and William Gascoigne are at the top of the list, along with that of William's uncle Sir Ferdinando Leigh.[20] Although the Protestation Oath had been intended to identify Catholics, it seems clear that many of them just signed it anyway. Few were as bold as an individual from the nearby village of Osset, whose entry on the record says: 'William Pasley a recusant within this parish and constablerie above said hath wilfully refused this protestation And further says that before he take it he

[18] Cf. Chapman, Allan, *Three North Country Astronomers* (Farnworth, 1982), 17, where it is stated (without reference to a source) that Gascoigne returned to Yorkshire in 1633.

[19] Foster, Joseph, *Op.Cit.*, John Gascoigne matriculated on 16th September 1634.

[20] House of Lords Record Office, *Protestation Returns (130), Agbrigg Wapentake*. A photocopy is held at the office of the West Yorkshire Archive Service in Wakefield, ref. Z125/1(L), 187.

will loose his life'. The most we can conclude is that, if William Gascoigne and his immediate family were Catholics, they were not vocal about it.

Maybe William, at least measured against the conventions of the time, was not all that interested in doctrinal issues of religion. His preoccupations were of another kind. After moving back to Middleton, it soon became clear that he did not fit the traditional mould of the local gentry. Whereas others spent their leisure time engaged in hawking and similar pursuits, William was captivated by the study of the natural world and the world of instruments or machines. And no doubt this interest was cultivated, maybe even inspired, by his father—Gascoigne senior was reputed to have a barn full of mechanical devices. According to Christopher Towneley, Henry Gascoigne was 'much given to Mechanick by which hee hoped to performe no meane thinges'.[21]

William Gascoigne became increasingly absorbed by astronomy and optics—'a late intruder into these studies' as he characterised himself. Though 'late' in his own estimation, nevertheless he must have started these studies well before his mid-twenties, for by the winter of 1638 he was already a seasoned and dedicated astronomical observer. Nevertheless, much of his time was spent perforce on other duties. He still considered himself very much a part-time astronomer, declaring to Oughtred 'I bestow only part of that time in these studies which other gentlemen, our neighbours, spend in hunting.'[22]

Remarkably, despite his school and university education, whether or not free from religious constraints, by the time Gascoigne arrived in Middleton, he had still not been taught any mathematics, 'having never had so much aid as to be taught addition'. Nor did he know 'what any proposition in Geometry meant'. But these difficulties didn't dampen his enthusiasm for learning. He started to instruct himself in the rudiments of mathematics and natural philosophy. In lieu of a teacher he used books, such as Oughtred's *Clavis*, and immersed himself in the latest writings on optical science.

[21]The National Archives, RGO 1/40, f.22r.

[22]Gascoigne to Oughtred (2/Dec/1640), MSS. Add. 9597/13/5/103, r (transcribed in Rigaud, Op. Cit., v.1, 34).

Chapter 9
The Optical World of William Gascoigne

During Gascoigne's lifetime the science of optics was in its infancy. Great strides had been made in the Middle Ages by arab scholars, such as Ibn Is-haq (al-Kindi) and Ibn al-Haitham (Alhazen), in understanding the nature of vision, but at the dawn of the seventeenth century the subject of lenses and mirrors was still largely the domain of craftsmen. The making of lenses for spectacles—progressing through trial and error—was the subject of family secrets. When Galileo turned a telescope to the heavens in 1610, however, discovering the craters and mountains of the Moon, the satellites of Jupiter, and the crescent phases of Venus, everything changed. The secrecy was still present—indeed, not until 1638, as his eyesight began to fail, did Galileo privately reveal (to Mariani) his own method of grinding and polishing lenses—but the quest to understand the behaviour of optical systems gained a new impetus. *Dioptrics*—the theory of lens-based image formation—and *catoptrics*—that of mirror-based systems—began to make real progress.

Even so, very few books were to be had that explained the art of telescope making and lens grinding. Those published were by scholars, rather than the secretive practitioners. Gascoigne pored over whatever books he could lay his hands on: Books that straddled the medieval optical world and the era of the telescope. To read them one had to grapple with optical concepts that still had vestiges of ancient suppositions.

Idols and Species

The notion of Plato (c.428–347 BC), Aristotle (384–322 BC) and Euclid (fl. c.295 BC) that vision was based on the transmission of 'visual rays' or 'ocular beams' from the eye, which reached out and lightly touched distant objects, had long since been abandoned. The journey to a more realistic theory, however, was a long one. Along the way, a popular theory, articulated most clearly by the Epicurean poet Lucretius (c.99–c.55 BC) and persisting until the Middle Ages, was that vision was caused by visible objects emitting a constant stream of simulacra, or thin films in the

D. Sellers, *In Search of William Gascoigne: Seventeenth Century Astronomer*, Astrophysics and Space Science Library 390, DOI 10.1007/978-1-4614-4097-0_9,

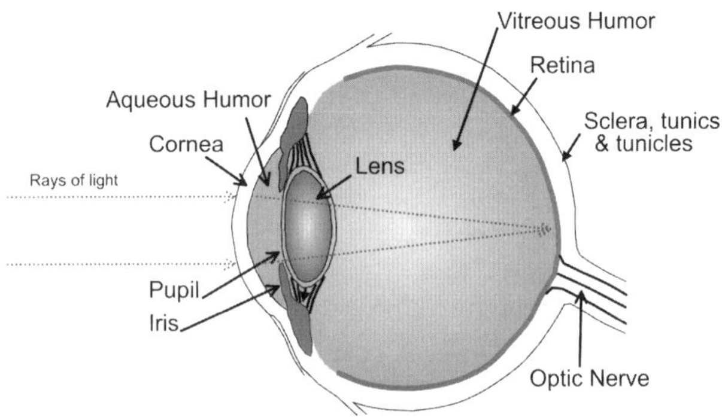

Fig. 9.1 The anatomy of the eye

form of the objects, which impinge upon the eye (Fig. 9.1) and provoke vision. Epicurus called these *eidola*, or idols. They were supposed to be somewhat like the skin discarded by a snake. 'Idols', wrote Lucretius, 'are sort of membranes stripped from the surfaces of objects, and float this way and that through the air ... each bears the appearance of the object from whose body it is shed'.[1] The term *idol* later gave way to the term *species*, to describe that which acts on the eye to cause vision. The mismatch between the observed object and the size of the pupil, admitting the species into the eye, was not considered to be an insuperable problem. The species were thought to shrink in size as they made their way from the emitting object towards the observer. Gascoigne, in his correspondence with Crabtree, Digby and Oughtred uses the terms *idol* and *species*, but clearly has a somewhat different take on their meaning. The lack of an evolved technical language is a significant stumbling block whenever new discoveries leap ahead of established concepts.

Al-Kindi (pre-800–c.866), living Baghdad, was steeped in the optical ideas of classical Greece. He accepted Euclid's theory that light travels in straight lines, but found it hard to swallow any notion of visual rays leaving the eye. Dragging optical science into a recognisably modern form, he argued that vision took place as a result of rays striking the eye and producing a sensation.

The central figure in the study of optics in the Arabic world was Ibn al-Haitham (c.965–c.1040), a native of Basra, working in Egypt. His masterpiece, *Kitab al-Manazir* (Book of Optics), endured as an influential text for the best part of 500 years. Eschewing reliance on previous authorities, it depended instead upon experiment and mathematics for its conclusions. Al-Haitham carried out practical experiments concerning the phenomena of luminosity, transparency and reflection. He also studied refraction, the bending of light as it passes from one medium to another: the same phenomenon that makes a pencil, half-submerged in a tumbler of water, appear to be broken. He found that the angle through which a ray of light is bent, or refracted, varies according to the angle at which the ray strikes the boundary between the different media—the 'angle of incidence'. Moreover, he correctly concluded

[1]Lucretius, *On the Nature of Things* (De Rerum Natura), book 4, lines 29–58, translated by Smith, Martin F., (London, 1969), 130.

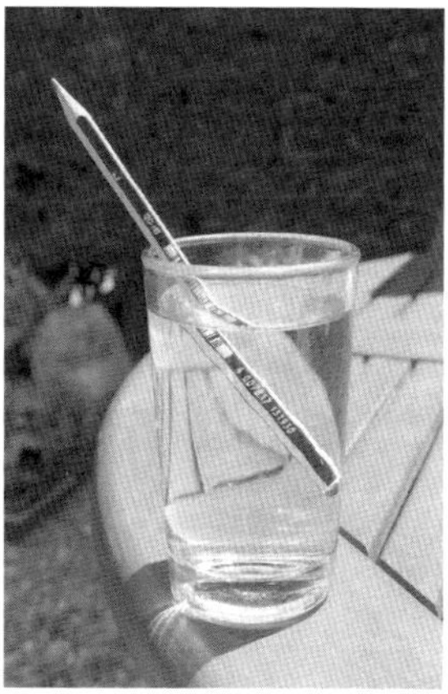

Fig. 9.2 Refraction in a tumbler of water

Fig. 9.3 Emission of rays from a point

that the ratio between the angles of incidence and refraction does not remain constant as the angles change (Fig. 9.2).

Starting from al-Kindi's supposition (that vision is due to rays striking the eye from without), elaborating a more complete explanation of vision, al-Haitham taught that every point on an object emits images of itself in all directions (Fig. 9.3): 'from each point of every coloured body, illuminated by any light, issues light and colour along every straight line that can be drawn from that point.'[2] Furthermore,

> through each point on the surface of the eye pass simultaneously the forms of all points in the visual field, but the form of only one point passes directly [i.e. without refraction] through the transparency of the tunics of the eye, and that is the point located at the extremity of the perpendicular issuing from the point on the surface of the eye. The forms of all the remaining points [in the visual field] are refracted at that point on the surface of the eye and pass through the transparency of the tunics of the eye along oblique lines.[3]

In other words, the unrefracted rays—having met the curved surface of the eye in a perpendicular fashion—form a cone of light pointing to the interior of the eye.

[2]Lindberg, David C., *Theories of Vision from Al-Kindi to Kepler* (Chicago and London, 1981), 73. Quoting from Alhazen's *De Aspectibus*, bk.1, Chap. 5, Sect.19, 10.

[3]*Ibid*, p.74 (Alhazen's *De Aspectibus*, bk.1, Chap. 5, Sect.18, 9.

Keeping the same arrangement, relative to each other, as they had when they left the emitting object, they reproduce, according to al-Haitham, a miniaturised image of that object. The theory was flawed, the refracted rays could not so easily be left out of the explanation, but it was a major step forward and managed, elegantly, to demonstrate how the likeness of a large object could be introduced through the tiny aperture of the eye, without resort to shrinking idols.

Al-Haitham's treatise on vision, in common with other Arabic masterpieces was eventually translated into Latin. As *De Aspectibus*, it thus became known to thirteenth century Europeans such as the English writer Roger Bacon (c.1214–1294) and the Polish theorist, Witelo (c.1235–post 1281).

Kepler and the Retinal Image

Witelo brought al-Haitham to a wider audience of natural philosophers through his magnum opus on optics, *Perspectiva* (written between 1270 and 1278, though the first printed version didn't appear until 1535). This book remained the definitive work on the subject for nearly 350 years. Ironically, its demise and the advent of modern geometrical optics were brought about by a work which purported to be simply a supplement to the thirteenth century text. In 1604 Johannes Kepler (1571–1630) published *Ad Vitellionem Paralipomena* (Supplement to Witelo). In it he introduced the first correct account of image formation.

Witelo and his predecessors, following in the footsteps of al-Haitham, had found it difficult to understand how every point on an object could emit rays in every direction and yet the rays entering the eye managed to produce a distinct—not blurred—image. They got around the problem by assuming that only the rays falling perpendicularly on the eye's surface had a role in producing the image: just one ray for each point on the object. Now, Kepler introduced a bold new concept: All the rays from such a point, which met the surface of the eye, were brought to a focus in a corresponding part of the image. What is more, the image was formed not in the middle of the eye or at the lens: It was formed at the back of the eye, on the retina, and it was reversed, both laterally and vertically. The reversal was a necessary consequence of the way the collection of rays from points at each side of the field of view were brought together at their respective focal points, as they were differentially refracted (depending on their angle of incidence) through the cornea, the lens and the vitreous humour.

> I say that vision occurs when an image of the whole hemisphere of the world that is before the eye, and a little more, is set up at the white wall, tinged with red, of the concave surface of the retina. How this image or picture is joined together with the visual spirits that reside in the retina and in the nerve, and whether it is arraigned within by the spirits into the caverns of the cerebrum to the tribunal of the soul or of the visual faculty … I leave to the natural philosophers to argue about.[4]

[4] Kepler, Johannes, *Optics: Paralipomena to Witelo, & Optical Part of Astronomy/* translated by W. Donahue (Santa Fe, 2000), 180.

To demonstrate his explanation of the workings of the eye, Kepler relied upon urinary vessels filled with clear water. With one of these he suggested, within a relatively darkened room, one could project a perfectly sharp inverted image of an illuminated window onto a piece of white paper, in much the same way that the human eye forms images on the retina.

Telescopes

The world of optics arrived at a watershed in the year 1610, 2 years before the birth of Gascoigne, when Galileo Galilei (1564–1642) published an account of his astronomical observations with a telescope. Scarcely 10 months earlier, he had heard about the existence of a 'spyglass ... made by a certain Dutchman by means of which visible objects, although far removed from the eye of the observer, were distinctly perceived as though nearby'. In his account—*Siderius Nuncius* (Starry Message)—he admitted that this news had caused him to put aside all other matters and 'apply myself totally to investigating the principles and figuring out the means by which I might arrive at the invention of a similar instrument, which I achieved shortly afterward on the basis of the science of refraction'. The result was 'a lead tube in whose ends I fitted two glasses, both plane on one side while the other side of one was spherically convex and of the other concave. Then, applying my eye to the concave glass, I saw objects satisfactorily large and close.'[5] As for a detailed explanation of the principles that he had investigated, Galileo simply said 'on another occasion we shall publish a complete theory of this instrument': A promise that was never fulfilled.

The task of providing a theoretical explanation of the basic principles governing the working of the telescope fell to Johannes Kepler—at that time living in Prague. Kepler had received a copy of *Siderius Nuncius* in April 1610 and immediately recognised its importance. He had to wait until August of the same year before he was able to look through a Galilean telescope himself and see with satisfaction the moons of Jupiter. Throughout August and September he then worked assiduously on a manuscript explaining how rays of light ray pass through systems of lenses—converging through the convex ones and diverging through the concave. He was armed with only an approximate law of refraction, assuming that the ratio between the angles of incidence and refraction is constant. From his own experiments he gauged the constant to be approximately equal to 4/3 for an air-to-water transition. Because the angles he was dealing with were relatively small, his explanations were reasonably correct. The resulting book, *Dioptrice* (Dioptrics), was published in 1611. Amongst its important conclusions was the proposal that a different form of telescope was possible. In place of Galileo's concave (or diverging) eyepiece, a convex (or converging) eyepiece could be used, giving an inverted image (Figs. 9.4 and 9.5).

The upside down image was maybe the reason why most astronomers at first steered clear of the Keplerian format of telescope, even though usually for

[5]Galilei, Galileo, *Siderius Nuncius* (1610), translated by Albert van Helden (Chicago, 1989), 36–7.

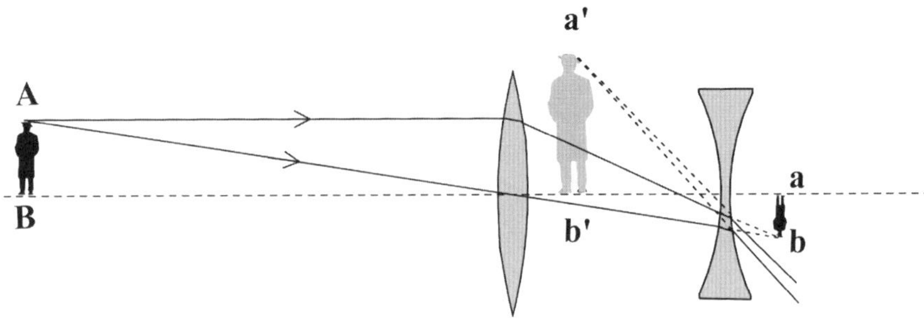

Fig. 9.4 The Galilean telescope. The observed object at AB gives rise to a virtual image at ab and an upright magnified apparent image at a′b′

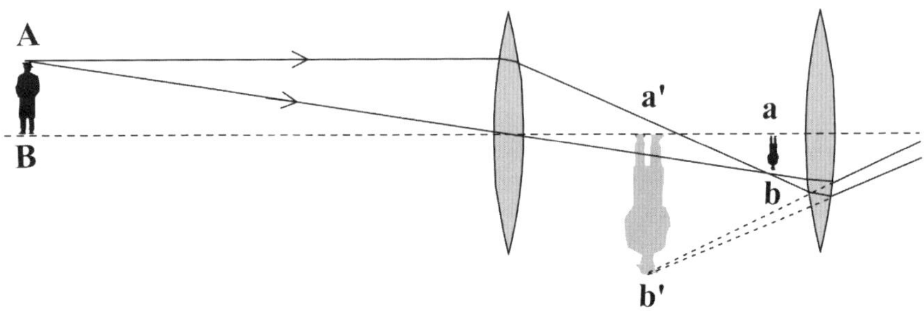

Fig. 9.5 The Keplerian telescope. The observed object at AB gives rise to a real image at ab and an inverted magnified apparent image at a′b′

observations of a celestial body, with the possible exception of the moon, inversion is of no consequence. Not until the 1640s—more than 30 years after the first description by Kepler—did its use become widespread. By then two distinct advantages of Kepler's design were becoming clear: Firstly, it had a markedly wider field of view—it could encompass a greater part of the vista before it, whilst still giving high magnification; and secondly, it provided a positive focus—a 'real' image—within the telescope tube.[6] The latter feature was indispensable for Gascoigne's invention of the telescopic sight.

Solar Projection

As with the Galilean telescope, the Keplerian version could also be used to project the image of a bright object onto a white screen or piece of paper. This was a particularly useful way to view the Sun: Indeed, the only safe way to observe the Sun without the risk of damage to the eyes. In this application, the Keplerian telescope

[6]Van Helden, Albert, *Measuring the Universe: cosmic dimensions from Aristarchus to Halley* (Chicago & London, 1985), 118–9.

had the advantage of creating an erect image on the screen, whereas the Galilean alternative gave an inverted image.

This is what attracted the Jesuit astronomer, Father Christoph Scheiner (c.1573–1650) to the Keplerian telescope. Scheiner was one of the earliest students of sunspots—those dark blemishes obscuring the dazzling face of the Sun that only came within reach of serious examination after the invention of the telescope. As early as 1611, he could be found in the belfry of the Holy Cross Church in Ingolstadt (Germany), carefully drawing the constantly changing disposition of the spots as they drifted roughly east to west across the Sun. In point of fact, Scheiner was initially doubtful that sunspots were truly stains on the Sun's orb: He considered it more likely that they were the silhouettes of small planets orbiting the Sun.

As soon as Scheiner attempted to keep a careful record of the position of the spots, he encountered a complication: The apparent shape of the Sun was not a perfectly circular disk. The Sun's image—particularly near sunrise or sunset—was a flattened, elliptical shape. In his book *Sol ellipticus* (1615) Scheiner correctly explained than this phenomenon was due to refraction of sunbeams through the Earth's atmosphere. Light from the lower limb of the Sun, striking the outer atmosphere obliquely, and passing through a thicker layer of atmosphere than that from the upper limb, would be refracted more strongly. Thus the lower limb would acquire a greater increase of its apparent elevation in the sky than did the upper limb and the vertical 'diameter' of the Sun therefore would be contracted. By contrast, the horizontal 'diameter' was unaffected. Thereby, the Sun appeared to be flattened. In a further work, *Refractiones coelestes* (1617) he developed this explanation further and gave details of his instrument for projecting the sun's image (Fig. 9.6).

In addition to being a skilled mathematician, Scheiner had a flair for invention. Indeed, even before he arrived at Ingolstadt to commence theological studies, he was already celebrated for his invention of the pantograph—a device for copying and enlarging drawings. Accurate recording of the sunspots placed new demands on his ingenuity. Improved mountings were required and in response to this challenge he devised the helioscope—an equatorially mounted telescope for projecting the image of the Sun. He was also acutely aware of the shortcomings of the available optical components: 'A clear, transparent, pure, homogeneous and even lens is considered equal to a gemstone!'[7]

Although Scheiner was under pressure to make his astronomical opinions conform to the prevailing doctrine of the Church, in optical theory he could let his ideas roam wherever experiment took him. And Scheiner was a great believer in experiment: 'A single true observation', he declared, 'ridicules a 1,000 hair-splitting arguments'.[8] His book *Oculus Hoc Est* (This is the Eye), published in 1619, embraced this outlook as it delved into the physiology and optics of the eye. Here he showed

[7]Scheiner, Christoph, *Rosa Ursina sive Sol* (Bracciano, 1626–1630), Book 2, Chap. 19, 97/II/40, quoted in *The Physicist and Astronomer Christopher Scheiner: Biography, Letters, Works*, Franz Daxecker (Innsbruck, 2004), 130.

[8]Scheiner, Christoph, *Op.Cit.*, Book 4, Chap. 33, 472/II/21, quoted in Franz Daxecker, *Op. Cit.*, 139.

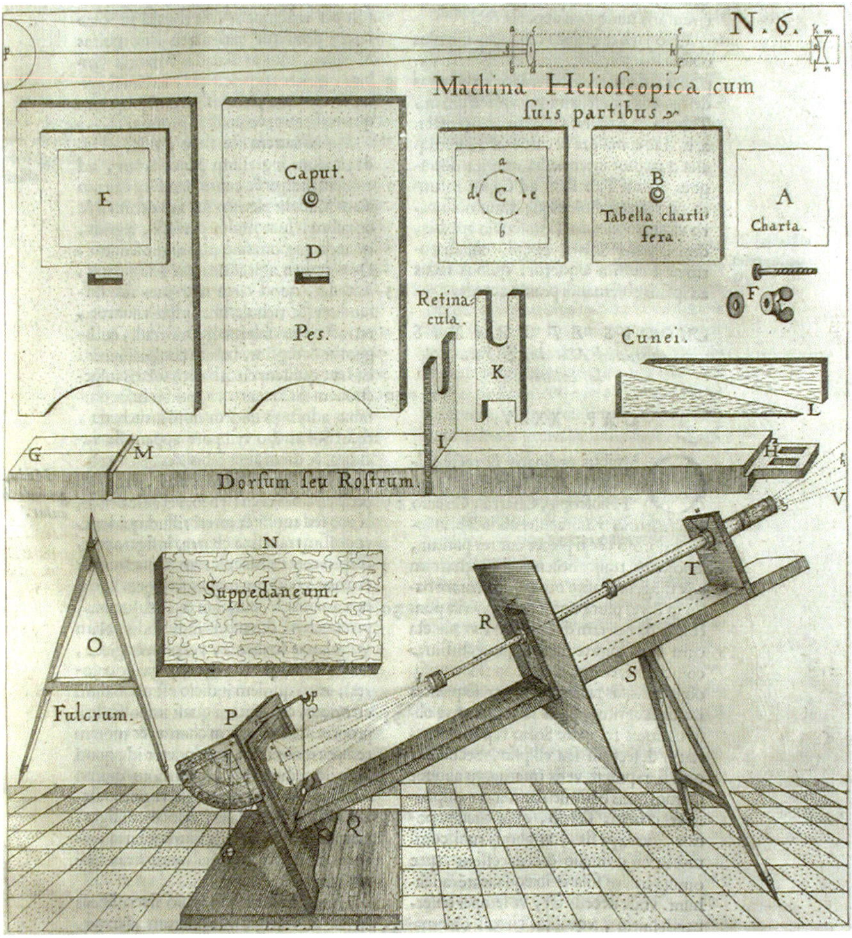

Fig. 9.6 Scheiner's instrument for projecting an image of the Sun (Image courtesy of History of Science Collections, University of Oklahoma Libraries)

an ingenious method of working out the curvature of the cornea—by putting glass spheres of varying radii next to the eye, until both sphere and eye showed an equal-sized reflection of a window, at which point the curvatures were deemed to be equal. The inner structure of the eye was determined by dissecting bovine eyes and the refractive index of the fluids within the eye was measured by placing samples in small spherical phials or bottles, which were then immersed in water to see how much small objects placed behind them would be magnified: An experiment that Gascoigne would later repeat, but with a typically ingenious twist.

Scheiner also proved that light rays entering a camera obscura[9] through a pinhole intersect between the pinhole and a screen to produce an inverted image on the

[9] *camera obscura* (Latin: a darkened room).

screen. In his subsequent book, *Rosa Ursina* (1630), his direct observation of the same intersection within the eye was recounted:

> I have proved, I may add, that the rays intersect before the image of the object is projected on the retina. Not only have I showed this in my work *Oculus* with many plausible experiments and thoughts, I also saw clearly when looking at the human eye in Rome in the year of the Lord, 1625, that the rays of candlelight entering through the pupil crossed before touching the retina when the sclera had been removed. I had conducted this experiment with numerous animal eyes.[10]

But in all his expositions of the effects of refraction—whether concerning flattening of the Sun's disk, bending of light through the lenses of telescopes, or focusing of rays within the eye—Scheiner was only equipped with a qualitative account of the effect.

The Law of Refraction

The precise law of refraction—the relationship between the angle of incidence of a ray of light on a boundary between different media, say air and glass, and the angle of the refracted ray—had intrigued astronomers and other investigators for almost one and a half thousand years. Ptolemy (c.90–c.168 CE) established that the path of an incident and refracted ray was the same, whichever direction it was travelling in between different media. He also carried out a series of experiments with light travelling between air and water, air and glass, and water and glass. These showed that the relationship was not a simple proportionality.[11]

The correct law was outlined in a treatise called *On the Burning Instruments* written around 984 by the Baghdad scholar Ibn-Sahl. Unfortunately, his treatise was unknown in Europe and the manuscript was not rediscovered until the 1990s.[12]

Robert Grosseteste (c.1168–1253), the Bishop of Lincoln, guessed that the angle of refraction was about half of the angle of incidence. Kepler, as we have noted, thought that the ratio between the two angles was constant and equal to about 4/3. It fell to the Dutch mathematician, surveyor and astronomer, Willebrord Snellius (Snel, 1580–1626) to announce the correct law of refraction. Snellius—commonly known as Snel—had been spurred into the public debate of astronomical questions by reading the first treatise of Scheiner on the sunspots. In 1612 he published anonymously *De maculis in sole animadversis, et, tanquam ab Apelle, in tabula spectandum in publica luce expositis, Batavi Disertatiuncula* ('A small treatise of a Dutchman about the spots that have been observed on the Sun and that have been

---

[10] Scheiner, Christoph, *Op.Cit.*, Book 2, Chap. 23, 110/II/39, quoted in Franz Daxecker, *Op.Cit.*, 131.

[11] Smith, Mark A., *Ptolemy and the foundations of ancient mathematical optics: a source based guided study* (Philadelphia, 1999), 144.

[12] Al-Khalili, Jim, *Pathfinders: The Golden Age of Arabic Science* (London, 2010), 156–8.

Fig. 9.7 René Descartes (Courtesy of the Library of Congress)

publicly exhibited, on view in a painting like the one made by Apelles'), criticising some of Scheiner's earliest conclusions.

In 1621, just a few years before his death, Snellius announced his finding that, for a given combination of media, it was not the ratio of the angles of incidence and refraction that was constant, but the ratio of their sines. Although this conclusion was not published in his lifetime, through correspondence it quickly circulated amongst students of optics. Papers belonging to the English astronomer Thomas Harriot show that he too had discovered this law, as early as 1601, but had not communicated it to anyone. Not until 1637 was the sine law published. It then appeared in the fifth volume of Pierre Hérigone's *Cursus Mathematicus* and, more famously, in the Second Discourse of René Descartes' *La Dioptrique* (Optics) (Fig. 9.7).

Descartes (1596–1650) was born at La Haye, in Touraine, France, into an aristocratic family. After spending his early manhood soldiering and travelling, in his mid-twenties he was inspired to turn to a secluded life of intellectual pursuits. He is remembered most widely for his contributions to philosophy ('I think, therefore I am') and analytical geometry (spatial relationships expressed in terms of x and y coordinates). However, he also made important contributions to optical science. *La Dioptrique* (1637) gave a clear description of the sine law of refraction, albeit in a slightly unfamiliar form. Referring to Fig. 9.8, he explained,

> the ratio or proportion between these angles [of incidence, ***i***, and refraction, ***r***] varies with all the different inclinations of the rays; whereas that between the lines AH and IG or others such, remains the same in all refractions caused by the same bodies.[13]

This was exactly equivalent to the modern formulation of the law, viz, that the ratio of the sine of the incident angle, ***i*** (ABH) to the sine of the refracted angle ***r*** (IBG) is constant.

Using this law, in *De l'arc en soleil (*On the rainbow), he gave the first correct explanation of the formation, size and shape of the rainbow. He also showed,

[13] Descartes, René, *Discourse on Method, Optics, Geometry, and Meteorology*, trans. and ed. Paul J. Olscamp (Indianapolis, 2001), 80–1.

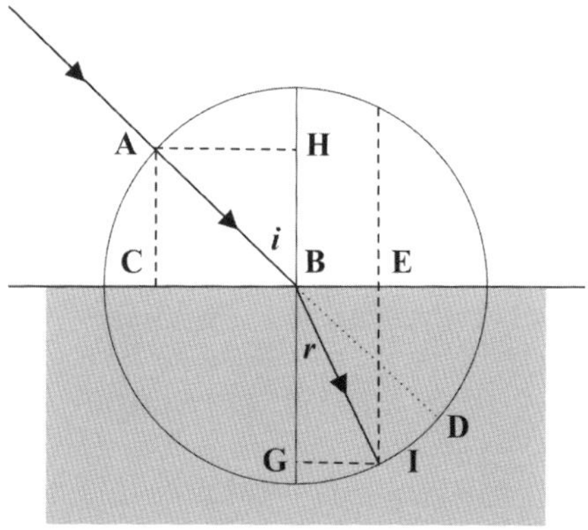

Fig. 9.8 Descartes' diagram of refraction

elsewhere in his work, how spherical lens surfaces were inherently subject to the problem of spherical aberration: An inability to focus the outermost rays of a light beam at the same point as the innermost rays. His solution to the problem—an aspherical lens—unfortunately was beyond the technology of his time.

In 1649, Descartes accepted an invitation from Queen Christina of Sweden to join her court and to instruct her in philosophy. But, far from finding there a comfortable abode to continue his academic pursuits, the French philosopher found the dynamic young queen to be a harsh taskmistress. She insisted that lessons commence at 5 a.m. in a bitterly cold library. Descartes' constitution was not up to the rigours of this new lifestyle and he contracted an inflammation of the lungs, which led to his premature death a few months later, on 11th February 1650.

The sine law of refraction, despite being published by Descartes in 1637, took a long time to be generally accepted or even widely known. As late as 1663, in his celebrated book *Optica Promota*, the Scottish mathematician James Gregory seemed to be unaware of the law.[14]

Gascoigne's Experiments

William Gascoigne, by contrast, was an earnest student of Descartes' work—in particular *La Dioptrique* and *De l'arc en soleil*—mentioning the French author's name no less than five times in his second letter to William Oughtred. He carried out his own experiments to confirm the sine law of refraction and to measure the refracting capabilities of different glasses. Going further than Descartes, he showed how to apply the sine law to the passage of sunrays through a Galilean telescope in order to calculate, from the size of the projected image of the sun, the apparent solar diameter. He explained the procedure in letters to Crabtree and Oughtred, using as an

[14]Diijksterhuis, F.J., *Lenses and Waves: Christiaan Huygens and the Mathematical Science of Optics in the Seventeenth Century* (Dordrecht, 2004), 38.

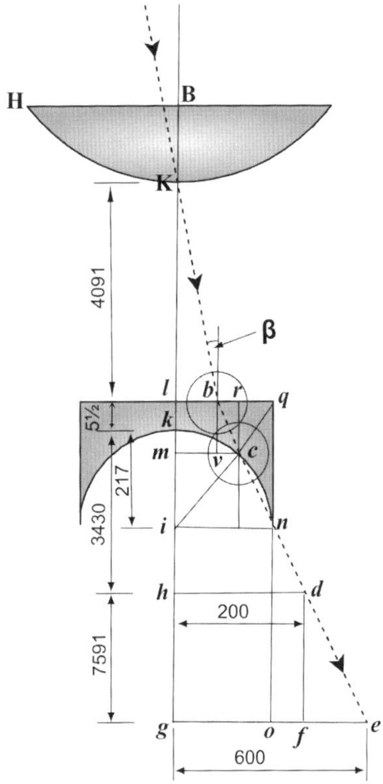

Fig. 9.9 Gascoigne's solar projection diagram (dimensions are in hundredths of an inch)

example an observation of the Sun made on 31 October 1640. The lenses, he confesses, are not 'according to my desire, yet they will as sufficiently as most other London best sale glasses serve for trials'.[15]

The image of the Sun was projected approximately 2.8 m beyond the eyepiece of the metre-long telescope to form a 300 mm diameter illuminated disc on a screen (see Fig. 9.9). This arrangement probably used a support system similar to that used by Scheiner (Fig. 9.6). In Gascoigne's case, however, the projection distance was much longer. It also differed in that Gascoigne placed between the eyepiece and the screen an intermediate moveable board (at *hd*, about a third of the distance from the eyepiece) in the middle of which there was a large hole, allowing the beam of light to pass through. Straddling the hole was a removeable wooden cross, covered with paper, upon each limb of which Gascoigne had marked arcs from circles of set diameters: 1, 1½, and 2 in.. When required, this cross enabled the intermediate size of the Sun's image to be measured. The comparison of the sizes of the pair of images, together with their distance of separation, he then used to calculate the angle of spread of the beam leaving the eyepiece.

[15] Rigaud, Stephen J., *Correspondence of scientific men of the seventeenth century* (Oxford, 1841), i, 39.

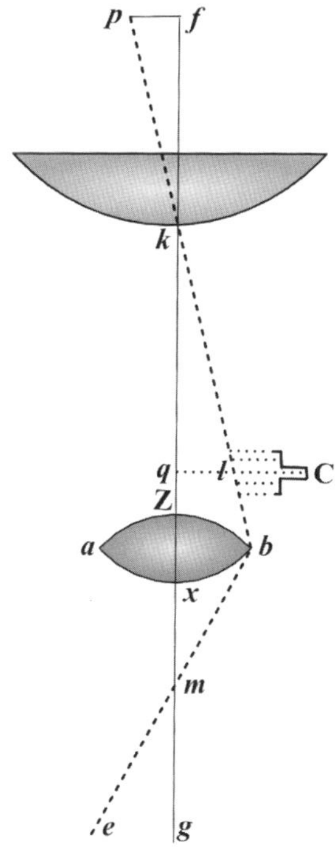

Fig. 9.10 Gascoigne's micrometer diagram

Armed with this angle and a knowledge of the refractive index of the eyepiece glass (1.537) derived from previous experiments, he was able to deduce the angle subtended by the Sun's semidiameter at the object glass. This he calculated to be 16′11″ on the date in question.

Once Gascoigne had his micrometer working satisfactorily, he was able to dispense entirely with such elaborate calculations based on the path of the sunrays through the concave eyepiece onto the screen. Instead, he could ascertain more-or-less directly the angle of the rays emerging from the object glass by measuring the size of the real image formed within the telescope against the moveable pointers (at *ql*) of the micrometer mounted on a Keplerian telescope (see Fig. 9.10).

Because the letter to Oughtred gave such a detailed description of his observations and apparatus—along with a record of the measurements—it has been possible for later enquirers to form a reliable opinion about the nature of his telescopes. And no one was a more assiduous enquirer than the Lancashire engraver, Sidney Gaythorpe (1880–1964). Despite being pre-occupied by the family business in Barrow-in-Furness that he was obliged to enter rather than going to University as he desired, Gaythorpe found time to carry on a single-minded 50-year pursuit unearthing information about Jeremiah Horrocks—a fellow Lancastrian, who as we have seen was the bosom friend and correspondent of Crabtree.

Gaythorpe had hoped to write a definitive work on Horrocks, but—in the words of the obituary published by the Royal Astronomical Society—'the unfortunate destruction of his house in 1941 during the last war and the consequent dislocation of his life, together with the need to remove his considerable library to a smaller house, delayed matters until ill-health overtook him and the project became unfeasible'.[16]

The pursuit of Horrocks inevitably brought Gaythorpe into contact with the correspondence of Gascoigne. Whereas Horrocks had written little in the way of a description of his observing instruments—characterising his transit of Venus telescope simply as one 'much more accurate than those generally used'[17]—Gaythorpe was intrigued to find that Gascoigne had in effect left a complete specification of the telescope used to project solar images. In a paper communicated to the British Astronomical Association *On a Galilean Telescope made about 1640 by William Gascoigne, Inventor of the Filar Micrometer*, Gaythorpe pointed out that

> The telescope itself has no doubt long since disappeared, but Gascoigne's measurements of the aperture, thickness, radius of curvature, and refractive index of the plano-convex object glass, as well as of the plano-concave eye-lens, are to be found in his second letter to William Oughtred .... From the two last quantities can be computed the focal length of each lens ...[18]

From these also Gaythorpe deduced that Gascoigne's telescope had a magnifying power of 11.2 times and a field of view of about 25′ of arc. Given the refractive index, he concluded that the lenses must have been made of light flint glass.

Correspondence with Crabtree that was unavailable to Gaythorpe has subsequently come to light, which also enables the lens characteristics of Gascoigne's micrometer-equipped Keplerian telescope to be determined.

Basing himself on the most advanced discoveries in optical science, the Yorkshire astronomer had arrived at the very frontier of instrumentation. He had developed devices and measurement techniques far superior to any in use elsewhere in the international scientific community: Devices which, as he knew, were essential for the next step in astronomical discovery.

[16]*Quarterly Journal of the Royal Astronomical Society*, v.6 (1965), no. 2, 248–9.

[17]Whatton, Arundell Blount, *Memoir of the life and labors of the Rev. Jeremiah Horrox* (London, 1859), 115.

[18]Gaythorpe, Sidney B., *Journal of the British Astronomical Association*, April 1929, 239. Gaythorpe summarised the characteristics of Gascoigne's Galilean telescope as follows:

	Object-glass (plano-convex)	Eye-lens (plano-concave)
Diameter (approximate)	2.25 in.	1.25 in.
Aperture (approximate)	1.20 in.	0.75 in.
Thickness (approximate)	0.105 in.	0.055 in.
Radius of curvature	25.8 in.	2.17 in.
Focal length	44.8 in.	–4.00 in.
Refractive index	1.576	1.542

Chapter 10
The Astronomical World of William Gascoigne

The prediction of planetary positions, distances in the solar system, the status of the Sun, and the age of the world: these were the big questions of astronomy during Gascoigne's lifetime. The alternative cosmologies of Ptolemy and Copernicus were still being hotly disputed. Kepler's elliptical orbits had been embraced by few English astronomers.

Of the *nature* of the stars and the planets themselves, nothing was known.

Following the invention of the astronomical telescope the disc-like appearance of the planets became evident and their relative size became an subject of study. The Sun and the Moon too became objects of close scrutiny.

As soon as Galileo (Fig. 10.1) turned his telescope to the skies in 1609–1610 he began to make remarkable discoveries that were at odds with previous orthodoxy. He discovered that the surface of the Moon was not perfectly smooth, but was covered with craters and mountains. He found that Venus had waxing and waning phases like those of the Moon—suggesting that it was illuminated by reflected light and indeed was orbiting the Sun rather than the Earth. He discovered that Jupiter was accompanied by four satellites, or moons of its own—thus demonstrating convincingly an example of celestial motion that was not centred on the Earth. By late 1611 he had also discovered that the Sun itself, far from being an unblemished shining orb, was tainted by dark blotches or spots. It was not clear whether these sunspots were features on the surface of the Sun or objects somewhere between the solar disk and the observer. One thing that was certain was that the spots were inconstant—changing their position on the surface of the Sun in a westerly direction, sometimes varying in shape and number, sometimes disappearing altogether (Fig. 10.2).

Galileo did not know that the spots had been previously seen (18 December 1610) in England by Thomas Harriot, because Harriot had not published his observations. The first known publication concerning the sighting of the spots using a telescope was by the Frisian[1] astronomer Johannes Fabricius in a 22-page pamphlet *De Maculis in Sole Observatis, et Apparente earum cum Sole Conversione Narratio*

[1] *Frisia*: a coastal region extending from the north-western Netherlands to the Danish border.

D. Sellers, *In Search of William Gascoigne: Seventeenth Century Astronomer*, Astrophysics and Space Science Library 390, DOI 10.1007/978-1-4614-4097-0_10,

Fig. 10.1 Galileo Galilei (Courtesy of Library of Congress)

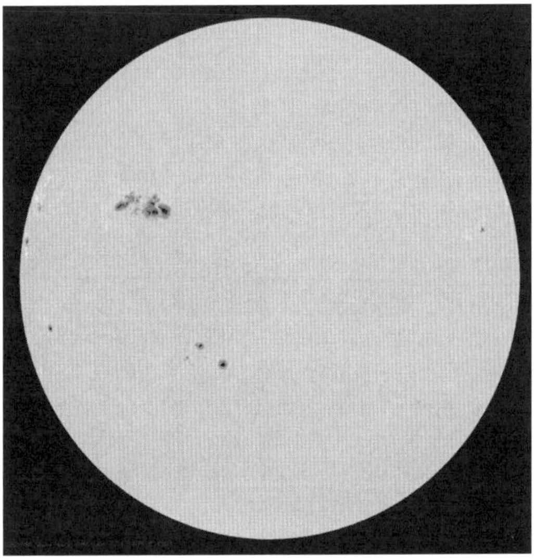

Fig. 10.2 Sunspots in November 2011 (NASA/Solar Dynamics Observatory)

(Narration on Spots Observed on the Sun and their Apparent Rotation with the Sun), printed in Wittenberg on 13 June 1611.

> While observing these things carefully, *wrote Fabricius*, a blackish spot suddenly presented itself, on one side indeed rather thin and faint, of no little size compared to the disk of the sun. I had at first no little doubt in the reliability of the observation, because a break in the clouds disclosed the rising sun to me, so that I thought that the clouds flying past gave the false impression of a spot on the sun. The observation was repeated perhaps ten times with Batavian telescopes of different sizes, until at last I was satisfied that the spot was not caused by the interposition of clouds.
>
> However, not willing to believe in the manifest testimony of my own eyes, on account of the strange and unusual appearance of the sun, I immediately called my father, at whose

Fig. 10.3 Christoph Scheiner (Courtesy of Stadtmuseum Ingolstadt)

> house I was then staying, having returned from Batavia,[2] in order that he might be present also to observe this [...] Thus the first day passed, and we left the sun, but not without great longing for its return on the morrow, so that our natural curiosity bore even the intervention of the night.
>
> Nevertheless we restrained our eagerness by anxious thought. For it was not yet certain whether that spot which we had seen would wait for the next observation.[3]

Fabricius was persuaded that the spots were on the Sun's surface: 'for what would we make of the spots if we did not place them on the sun itself I don't know'.[4]

In March 1611, the Jesuit priest Christoph Scheiner, along with his assistant Johann Cysat, noticed spots on the Sun as they looked at it through smoke rising from a fire that had broken out at the University of Ingolstadt (Fig. 10.3).

Between 12 November 1611 and 26 December 1611 Scheiner, responding to queries asked of Jesuit scholars in his area, wrote three letters to Marc Welser, magistrate and prefect of Augsburg, giving his views on the spots. At first, Scheiner

---

[2] *Batavia*: The Netherlands.

[3] Vaquero, J.M. and Vázquez, M., *The Sun Recorded through History* (London, 2009), 111.

[4] Reeves, E., and Van Helden, A., *On Sunspots – Galileo Galilei and Christoph Scheiner* (Chicago, 2010), 32.

suggested that the spots must be 'either on the sun or in some heaven outside the sun'. If *on* the Sun, then their translation across its disc would suggest that the Sun rotates. However, partly because the spots didn't seem to return to the same position, and partly because stains on such a 'lucid body' would be unfitting, he soon came to the view that the spots were small planets orbiting the Sun.

Shortly afterwards, the letters were published on Welser's private press, with the title *Tres Epistolae de Maculis Solaribus Scriptae ad Marcum Welserum* (Three letters about sun spots written to Marc Welser). Scheiner's superiors were fearful that open speculation about such a novelty might, if proved erroneous, bring the Jesuit order into disrepute or even ridicule. They forbade Scheiner to mention the order or the name of any Jesuit. Even his own identity had to be kept secret. As a result, the letters ended up signed with a pseudonym—'*Apelles hiding behind the canvas*': Apelles being the ancient Greek painter, who, according to Pliny the Elder, when he had finished his works would 'place them in a gallery in the view of passers by, and he himself stood out of sight behind the picture and listened to hear what faults were noticed'.[5]

The publication of the Three Letters from *Apelles* was to lead to a bitter priority dispute between Galileo and Scheiner over who saw the spots first: a dispute in which the language became more and more intemperate, with Scheiner losing no opportunity to condemn Galileo, and the latter ultimately calling Scheiner 'this pig, this malicious ass'! More importantly, the Italian astronomer disputed Scheiner's conclusion that the spots were tiny planets.

Galileo's first public pronouncement, in the Spring of 1612, similar to Scheiner's earliest comments, was cautious. He advised that the changing position of the spots showed 'either that the sun revolves in itself, or that perhaps other Stars, in like manner as Venus and Mercury, revolve about it'.[6] In January 1612, Welser wrote to Galileo asking for his opinion of Scheiner's Three Letters,

> You will do me a favour by freely telling me your opinion about these solar spots, whether you judge these substances to be stars or something else, where you believe they are situated, and what their motion is.[7]

Galileo responded by confirming Scheiner's view that the spots were undoubtedly real—not illusions or the result of optical defects—but, unlike Scheiner, he by now was convinced that the spots were on the Sun's surface:

> … the hypothesis that they cannot be on the solar body does not appear to me to have been fully and necessarily demonstrated. For it is not conclusive to say, as he does in the first argument, that because the solar body is very bright it is not credible that there are dark spots on it, because as long as no cloud or impurity whatsoever has been seen on it we have to designate it as most pure and most bright, but when it reveals itself to be partly impure

[5] Pliny the Elder, *Natural History*, tr. H Rackham (London & Cambridge Massachusetts, 1952), v.9, book xxxv, 84 (p.323).

[6] Reeves, E., and Van Helden, A., *Op.Cit.*, 78. The seventeenth century English of the translation by Thomas Salusbury has been updated.

[7] *Ibid.*, 87.

> and spotted, why shouldn't we call it both spotted and impure? Names and attributes must accommodate themselves to the essence of things, and not the essence to the names ...[8]

He disputed the notion that the spots are lightless silhouettes and maintained that they are intrinsically bright—just not as brilliant as other parts of the Sun—whereas Scheiner had said that the spots were much darker than any place on the Moon. Commenting on Scheiner's view that the fleeting nature of the spots showed that they could not therefore be counted part of its body, Galileo retorted:

> The argument would, however, be conclusive if he had first established that these spots were permanent, that is, that they were not arising anew and likewise fading away and vanishing. But whoever says that some come into being and others pass away will also be able to maintain that the Sun, turning on itself, carried them with itself without the necessity of presenting them to us unaltered, either arranged in the same order or with the same shape ... Apelles himself will have seen that some show themselves at their first appearance far from the circumference of the Sun, and others vanish and are lost before they finish traversing the Sun, for I, too, have seen many of these. I neither affirm nor deny, however, that they are on the Sun; I only say that it has not been sufficiently demonstrated that they are not on it.[9]

Refusing to let his judgement to be constrained by the limits of existing knowledge, he added,

> ... the substance of the spots could be a thousand things unknown and unimaginable to us.[10]

Galileo beseeched Welser to pass on his greetings to *Apelles*—whom, at this stage, he considered to be 'a person of sublime skill and a lover of truth', despite the difference of opinion. He promised to send some drawings of the spots 'of absolute precision, in their shapes as well as their daily changes in position, without a hairsbreadth of error, all made by a most exquisite method, discovered by one of my students' (Fig. 10.4).[11]

Accordingly, about 3 months later, Welser received a description of the telescopic projection method of solar observation—devised by the Benedictine monk, Benedetto Castelli, who was one of Galileo's students.

> in order to see very distinct and sharply delimited spots, it is best to darken the room, shuttering every window, so that no light enters it other than that which comes through the tube, or at least to make it as dim as possible and to fit a rather large piece of pasteboard to the tube such that it throws a shadow on the paper where one has to draw and prevents any sunlight from falling on it, except for what comes through the lenses of the tube. It should be noted next that the spots exit the tube inverted and located opposite to where they are on the Sun: that is, the spots on the right come out on the left side, and the higher ones lower, because the rays intersect each other inside the tube before they emerge from the concave glass.[12]

Eventually, Christoph Scheiner did come to accept the truth that the spots were on the face of the Sun itself and, indeed, in 1630 published a comprehensive study

---

[8] *Ibid.*, 91.

[9] *Ibid.*, 94.

[10] *Ibid.*, 98.

[11] *Ibid.*, 104.

[12] *Ibid.*, 126–7.

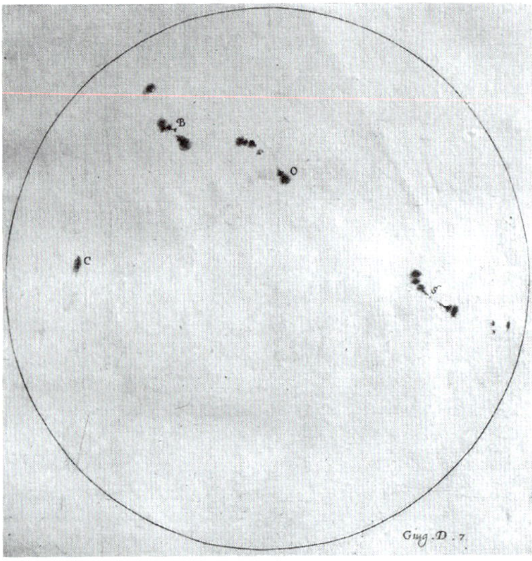

Fig. 10.4 Sunspot observations by Galileo in June 1612 (Image courtesy of History of Science Collections, University of Oklahoma Libraries)

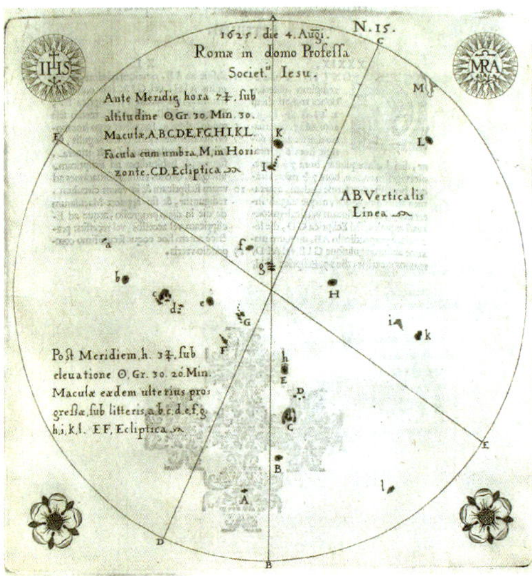

Fig. 10.5 Sunspot observations by Scheiner in August 1625 (Image courtesy of History of Science Collections, University of Oklahoma Libraries)

of their motions (and the implied solar rotation), *Rosa Ursina*, which became the standard work on the subject for well over a century (Fig. 10.5).

In *Rosa Ursina*, Scheiner explains how he too has adopted the telescopic projection method for solar observation, but he has realised the superiority of the keplerian telescope, with its convex eyepiece in lieu of the concave Galilean version. He knew from his study of Kepler's *Dioptrice* that this new configuration would conveniently produce an upright projected image (unlike Galileo's inverted one), but also remarked on another advantage.

> If you fit two like [convex] lenses in a tube … and apply your eye to it in the proper way, you will see any terrestrial object whatsoever in an inverted position but with an incredible magnitude, clarity and width. But also, you will compel any stars you wish to submit to your sight; for since they are all round, the inversion of the position of the total view is not confusing with respect to the visual configuration.[13]

He presented illustrations of a new apparatus for projecting sunspots that allowed them to be conveniently traced as the projection screen, along with the telescope tube, was moved to track the Sun in the sky. Scheiner's publication of this arrangement represented a landmark in observing procedure. From this time onward the keplerian telescope gradually displaced the Galilean one as the preferred 'astronomical' telescope.

Scheiner hadn't been alone in believing the spots to be tiny planets. The French astronomer, Jean Tarde (Tardius), a canon in the cathedral of Sarlat in southwestern France, went so far as to give a name to these 'planets'—*Borbonia sidera* (Bourbon stars) in honour of the French ruling dynasty. Not that Tarde and other clerical astronomers had a closed mind with respect to alternative theories about the nature of sunspots. Tarde was an ardent admirer of Galileo, whom he had visited in Florence in November 1614, no doubt hearing the case first-hand for assigning the spots to the surface of the Sun. Later on the same trip to Italy Tarde had discussed the spots in depth with the Jesuit astronomer Christopher Grienberger and his students in Rome. There he heard that 'the spots discovered on the star of the Sun were putting many people to a lot of trouble and that for some time they had been arguing about the composition, form, place, movement and duration of these spots. Some thought that they were a collection of small stars clustered together, very close to the sun, which go around it like Venus and Mercury or like the Sidera Medicea behind Jupiter. Others claimed that they are cavities in the solar body. Some thought they were perpetual, others said they had seen several disappear and pass away before managing to cross the face of the Sun'.[14] On balance, the canon from Sarlat plumped for the former interpretation.

William Crabtree was aware of Tarde's thesis from a description given by Libert Froidmund of Louvain, but was unimpressed. William Gascoigne, by contrast, was familiar with these works and with Scheiner's change of heart, but strangely—albeit with ingenious reasoning—he seems to have plumped for the position that Scheiner had abandoned. Hence, the fact that Crabtree went to great pains in his letter of 7th August 1640 to go through the same arguments that Galileo had put to a younger Scheiner.

> I must acknowledge you say more for the stellifying of these Solar obscurities [i.e. deeming them to be planets] than I have heard before; yet I conceive not sufficient, either demonstratively or probably to countermand those which Galilaeus, Kepler, and others have produced to the contrary.[15]

---

[13] *Ibid.*, 312.

[14] Tarde, Jean, *A la rencontre de Galilée: Deux voyages en Italie* (Geneva, 1984) (ed. F.Moreau and M.Tetel) (tr. DS), 79.

[15] *Philosophical transactions*, v.27, 288. William Crabtree to Gacoigne (7 Aug 1640).

If sunspots were a matter of contention, the rival solar system hypotheses—of Ptolemy, Copernicus and Kepler—were even more so. Gascoigne, Crabtree and Horrocks—pioneering new approaches of precision astronomy—were at the forefront of European attempts to determine which was correct.

In order to appreciate the character of the disputes, we need to contrast the alternative ideas.

From ancient times astronomers had believed that the motion of the night sky was clear evidence that the Earth was at the centre of the universe. Every night the blazing stars would slowly rotate from east to west. The Sun and the Moon too would complete the same journey.

Everything conspired to give the firm impression that the stars and planets were fixed to some heavenly vault that rotated about a stationary Earth. The sacred scriptures in the Bible and the teaching of the Christian Church reinforced this conviction. After all, did not Psalm 104, refer to 'God, who laid the foundations of the earth, that it should not be removed for ever'?

It was this earth-centred, or geocentric, view of the universe upon which the great astronomer of antiquity, Claudius Ptolemy (c.90–170 CE), had built the magnificent theoretical edifice in his book, *The Almagest*.

In Ptolemaic theory the Sun revolves in a circle around a static Earth in a plane inclined relative to the plane of the Equator. As a result the height reached by the midday Sun varies through the course of the year. As the northern hemisphere summer approaches, the Sun reaches a greater altitude with each successive day, until a point is reached when it seems to come to a standstill. This is the *summer solstice* and here we have the longest day combined with the shortest night. The Sun's peak altitude then diminishes with each passing day until the corresponding lowest point is reached (the *winter solstice*) with the shortest day and longest night.

In between the solstices there are two points when night and day are equal: the Vernal (Spring) and Autumn equinoxes (see Fig. 10.6). These key points punctuating the Sun's progress along its apparent annual path—the Vernal Equinox, the Summer Solstice, the Autumn Equinox and the Winter Solstice—marked the beginning of the seasons bearing the same names.

Although Ptolemy had the Sun moving in a circle about the Earth, it is crucial to note that the Earth was not placed at the centre of the circle. This was important for Ptolemy since it enabled him to account for the inequality in the lengths of the seasons. The lengths of the seasons measured by Hipparchus around 130 BC were[16]:

Spring	$94^1/_2$ days
Summer	$92^1/_2$ days
Autumn	$88^1/_8$ days
Winter	$90^1/_8$ days

[16] Evans, James, *The History and Practice of Ancient Astronomy* (Oxford, 1998), 210. Note that the lengths of the seasons are different in the modern era (currently, Spring, Summer, Autumn and Winter, are 93, 93, 90 and 89 days, respectively).

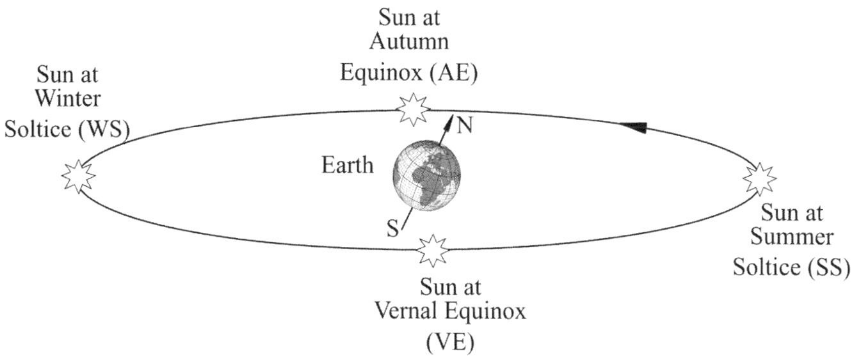

Fig. 10.6 The tilt of the Earth's rotational axis causing the seasons (in the geocentric model)

Clearly, the angular speed of the Sun was noticeably greater in December than in July—an inconstancy known as the *solar anomaly*. Since Ptolemy believed that in reality the Sun was travelling at a uniform speed in a perfect circle, then the only explanation for the greater angular speed in December was that the Sun was then nearer to the Earth. In other words, the Earth was offset from the *centre* of the circular orbit and the speed variation was the resulting, *purely visual*, effect. The amount of the offset was termed the *eccentricity* of the solar orbit. In theory, this eccentricity should result in the apparent diameter of the Sun varying throughout the course of the year, but such variation was far too slight to be detectable by Ptolemy and his contemporaries.

Ptolemy's view held sway for almost 1,400 years, until the magnum opus of the Polish cleric Nicolas Copernicus, *De Revolutionibus Orbium Coelestium*, emerged from the press in 1543—just as its author lay on his death bed. This work challenged the notion that the Earth was the stationary centre of the Universe. The hypothesis advanced in the new book was that the Sun was at the centre and the earth—a mere planet like Venus and Mars—was rotating daily on a north–south axis and revolving in an annual circular orbit about the Sun.

Many years would elapse before this heliocentric hypothesis met with universal accord. Initially, most astronomers rejected it outright. To them it seemed to fly in the face of common sense.

There were three major objections to Copernican theory. Firstly, if the apparent daily revolution of the stars was simply the visual effect of the Earth spinning on its axis once a day, it was clear that the surface of the Earth must be spinning in an easterly direction at a prodigious speed. Such a theory had been proposed by the Greek astronomer Aristarchus in ancient times and Ptolemy himself had disposed of it by saying that in such a case 'Neither clouds nor any other flying or thrown objects would ever be seen moving towards the east, since the Earth's motion towards the east would always outrun and overtake them'.[17]

[17] Ptolemy, C., *Almagest*, tr. GJ Toomer (Princeton, 1998), Book 1, Chap. 7, 45.

Secondly, if the Earth was travelling in a colossal orbit around the Sun once a year, then its speed must be immense. How could an object as massive as the Earth achieve this?

Thirdly, if the orbit of the Earth was so great, why didn't the stars have a different disposition when viewed from different sides of the orbit? Why, in other words, was there no discernible parallax effect—the simple effect that makes nearer objects seem to shift position relative to more distance objects when we view them first with one eye and then with the other eye?

The physics necessary for giving a convincing reply to the first two objections, however, was not available to sixteenth century astronomers. The answer to the third objection was that the terrestrial orbit was miniscule relative to the distance of the nearest stars and a parallax effect would only be detectable with the aid of instruments yet to be invented: descendants of Gascoigne's micrometer.

Although the heliocentrism of Copernicus offered a more elegant explanation of certain celestial phenomena, such as the retrograde motion of Mars, the problem was that its predictions of what could be seen in the sky differed little from those of the traditional Ptolemaic theories: The apparent motion of the Sun across the sky would be the same whether the Sun was in orbit about the Earth or vice-versa. Moreover, the explanation of the seasons was still essentially the same. In the Copernican system the constant tilt of the Earth's spin axis, relative to its orbital plane around the Sun, caused the seasons. And their varying lengths were reproduced by situating the centre of the Earth's circular orbit at a point offset from the centre of the Sun (Fig. 10.7), thus introducing an eccentricity akin to that of Ptolemy. Faced with these problems and considerations, most astronomers continued to believe in a geocentric system. Gascoigne's neighbourhood threw up a possible honourable exception: In the porch of St. Michael's Church in East Ardsley, less than 4 km from Gascoigne's house is the grave of local astronomer, John Feild (or, Field) (1520–1586). Field has been called 'the proto-Copernican of England'. In 1557 he published the first Almanac or Ephemeris in England avowedly based on 'Copernicus and Erasmus Reinholt, whose writings are established and founded on true, sure and plain demonstrations'.[18] Yet even Feild might not have held to a literal interpretation of the cosmology of Copernicus. Many astronomers—without necessarily accepting their heliocentric premises—did start to embrace *De Revolutionibus* and the *Prutenic Tables* (1551) of Reinhold at least as the most reliable recipe books for calculating planetary positions.

In 1609 Johannes Kepler, basing his analysis on the wealth of accurate observations produced by the Danish astronomer Tycho Brahe, published his book *Astronomia Nova* (New Astronomy), which demonstrated that the orbit of Mars is not a circle, but an ellipse in which the Sun is at one focus.

The ellipse had been familiar to geometers since antiquity. It is any closed curve in which, for every point on the curve, the sum of that point's distances from two fixed points is constant. Each of the fixed points is called a focus. The further the

[18] John Field, in *Ephemeris anni 1557 currentis, juxta Copernici et Reinhaldi canones, &c*, quoted by Rev Joseph Hunter, *Some particulars of the life of John Field, the proto-Copernican of England*, Gentleman's Magazine (London, 1834), v.1 (New Series), 493.

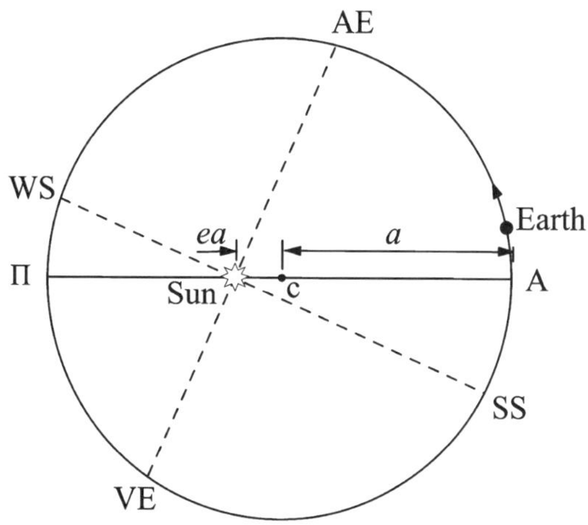

Fig. 10.7 The Copernican circular orbit of the Earth (The Earth's circular orbit has a radius *a* and is centred at *c*. The Sun is offset from the centre of the orbit by a distance *ea*, where *e* is the eccentricity. Points *Π* and *A* are the Earth's nearest and furthest points from the Sun—perihelion and aphelion. *AE*, *VE*, *SS*, and *WS* are the autumn & vernal equinoxes and the summer & winter solstices respectively)

focus is from the centre of an ellipse, then the more elongated or cigar-shaped the ellipse is: its eccentricity increases.

After deducing the shape of the orbit of Mars, Kepler then went on to derive a set of rules describing its rate of motion. Clearly, if Mars moves along an elliptical path, its distance from the Sun must vary with time. Kepler found that Mars moved faster when it was near to the Sun and slowed down as it reached the furthest point from the Sun. He found moreover, through painstaking calculation, that the area swept out by an imaginary line joining the planet to the Sun is always proportional to the time of travel. He subsequently proved that what applied to Mars also applied to the other known planets: Mercury, Venus, Earth, Jupiter, and Saturn (see Fig. 10.8).

In the Kepler arrangement, the seasons were still caused by the inclination of the Earth's axis and the effect that this had at different stages of the orbit. The inbuilt eccentricity of an elliptical orbit, however, coupled with the direction of that axis and the *real* variation in speed would suffice to give the seasons different lengths. Kepler was convinced that the Sun exerted a physical effect on the planets: It was the driving force of the solar system and planets, including the Earth, speeded up under its influence as they approached it. The net effect of the *real* speed increase was that Keplerian theory required only half of the Copernican eccentricity in order to explain the observed increase in angular speed. This *bisection of the eccentricity* was seen as a hallmark of the new theory.[19]

During Kepler's lifetime and the following few decades very few astronomers accepted—or even were aware of or understood—his rules for planetary motion.[20] Crabtree and Horrocks were startling exceptions, way ahead of their time. Horrocks, with unalloyed youthful enthusiasm, declared

---

[19] Wilson, Curtis *Astronomy from Kepler to Newton* (London, 1989), art.IX, 207 (originally in *Centaurus*, v.17 (1973), 207).

[20] For a detailed account of the reception of Kepler's work, see Applebaum, Wilbur, *Keplerian astronomy after Kepler: researches and problems*, History of Science, xxxiv (1996), 451–504.

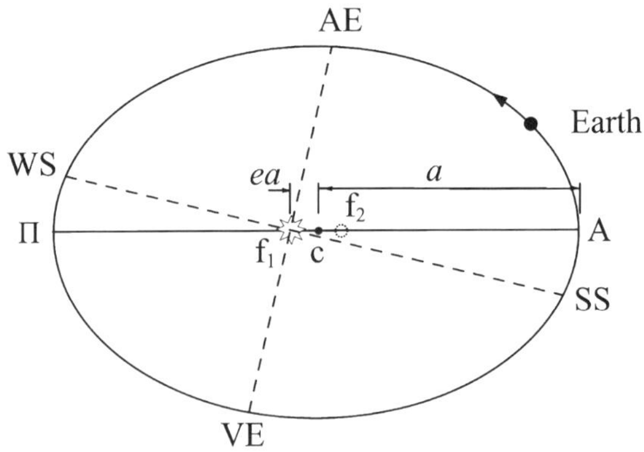

Fig. 10.8 The Keplerian elliptical orbit of the Earth (The Earth's elliptical orbit has a semi-major axis *a* and is centred at *c*. The Sun is offset from the centre of the orbit by a distance *ea*, where *e* is the eccentricity. Points *Π* and *A* are the Earth's nearest and furthest points from the Sun—perihelion and aphelion. *AE*, *VE*, *SS*, and *WS* are the autumn & vernal equinoxes and the summer & winter solstices respectively. Points f_1 and f_2 are the foci of the ellipse)

> that prince of astronomers, Kepler, to whose discoveries alone, all who understand the science will allow that we owe more than to those of any other person. I venerate with the greatest honour and admiration his sublime and enviably happy genius; and if necessary, I would defend with my best efforts the Uranian citadel of the noble hero who has so much surpassed his fellows, nor shall any one while I live, violate his ashes with impunity His death was an event that must ever have happened too soon.[21]

Predictive theories, constructed on the basis of the different hypotheses, came under scrutiny as the calculated planetary positions were compared with observations. The Rudolphine Tables (1627) of Kepler quickly demonstrated their superiority in predictive capability. The transit of Mercury in 1631 was to be particularly important in confirming this. The transit, on 7th Nov. 1631 (NS), was observed by Pierre Gassendi in Paris and revealed that Kepler's tables were out only 14′24″ in the predicted longitude of Mercury, whereas the Ptolemaic tables were out by 4°25′ and the Copernican ones by 5° (about ten moon-widths!). Each theory of course predicted a somewhat different eccentricity and shape of orbit for the Earth (or the Sun) and this was something that could also be scrutinised, though with much more difficulty. The key to effective assessment of these parameters was the fact that the apparent size of the Sun in the sky would change as its distance from the Earth varied. The ratio of the smallest and largest diameters could be used to determine the eccentricity and a series of diameter measurements through the year could enable the very shape of the orbit to be plotted. But an accurate value for the diameter of the Sun—let alone the subtle variation in that diameter throughout the year—was beyond astronomers .... Until, that is, Gascoigne's micrometer came along.

[21] Whatton, A.B., *Memoir of the Life and Labors of the Rev. Jeremiah Horrox* (London, 1859), 177.

Like Crabtree and Horrocks, Gascoigne first of all struggled to make sense of the much-vaunted tables of the Belgian astronomer, Philip van Lansberge (1561–1632). Lansberge had adapted the Copernican theory in such a way as to ensure agreement with ancient observations that he believed to be exact. Consequently he characterised his tables as perpetual—that is, valid for all time (*Tabulae motuum coelestium perpetuae*). His faith in the ancients was in stark contrast with Kepler, who frankly admitted that the Rudolphine Tables were often at odds with the tables and observations of Ptolemy. Lansberge thought this to be a scandalous abandonment of the true duty of the astronomer and both he and his co-worker Martinus Hortensius (1605–1639) became involved in bitter polemics against the adherents of Kepler.

Determining which system was superior—that of Lansberge or that of Kepler—was not a simple matter. It certainly could not be reduced to a single observation. After all, even the prediction of the 1631 transit of Mercury, which gave such a boost to Kepler's *Rudolphine Tables*, did not deliver a mortal wound to Lansberge's tables. As Hortensius pointed out, the *Tabulae Perpetuae* erred far less than the Ptolemaic, Copernican and Tychonic (Longomontanus) tables in their prediction of the transit (being out 1°8′ in longitude and 17′ in latitude). In his assessment of the transit, *De Mercurio in sole viso* (1634), Hortensius submitted that these other tables differed so much from the observation that 'no medicine would suffice to heal the wound'.[22] Admitting that the Lansberge prediction was not as good as Kepler's, he concluded 'therefore, after all the highly praiseworthy industry of the old man [i.e. Lansberge, who had died 2 years earlier at the age of 71], there no doubt remains something to be corrected in his theory of Mercury, although in all other observations the celestial truth is given accurately enough by his tables'. Probably echoing the views of many practically-minded astronomers, Hortensius spent some of his time ridiculing the harmonic speculations of Kepler. Posterity will laugh, he claimed, at the labour wasted on such idle dreams.[23]

It was true that Kepler (Fig. 10.9) had ploughed an inordinate amount of energy into studying semi-mystical harmonies in the motion of the planets. His first book, *Mysterium Cosmographicum*, had speculated illusory relationships between the sizes of the planetary orbits and the shapes of the five 'perfect' solids, which many astronomers rightly rejected. Even his masterwork *Harmonies of the World* was replete with attempts to discern musical analogies: so much so, that initially it was easy to miss, as one ploughed through it all, the nugget of gold which later became known as Kepler's Third Law—the proposition that the ratio between the square of a solar system planet's period of revolution (T) and the cube of its mean distance from the Sun (R) was constant (i.e., $T^2/R^3 = \text{constant}$, where T is in Earth-years and R is a multiple of the mean Earth-Sun distance).

By January 1641 Gascoigne had still not seen a copy of Kepler's *Harmonies of the World* and was asking Crabtree to get him a copy, provided, he pointedly added,

---

[22] Wilson, Curtis, *General History of Astronomy* (Cambridge, 1989) v.2A, 165.

[23] *Ibid.*, 165.

Fig. 10.9 Johannes Kepler (From Great Astronomers, by Robert Ball, 1895)

'it have anything in it but merely speculation'.[24] Even so, he was favourably predisposed towards Kepler, referring to him in the same letter as 'the excellent Kepler' and going on to quote from books such as *Paralipomena* and *The Epitome of Copernican Astronomy* by the German astronomer.

The pioneering approach of Gascoigne and his friends demanded that nothing be taken for granted in the work of the giants of astronomy—everything had to be tested or reassessed. Accordingly, he and Crabtree ploughed through dry scholarly texts on the chronology of the world and revisited the ancient equinox observations, which were at odds with Kepler, but agreement with which Lansberge deemed to be a merit of his own tables.

Crabtree rapidly lost faith in Lansberge and, early in his correspondence with Gascoigne, bluntly explained his reasons:

> *Lansberg* in Eclipses, especially of the Moon, comes often nearer the truth than *Kepler,* yet it is by packing together Errors; his Diameters of the Moon and Sun being false, and his variation of the Shadow being quite repugnant to Geometrical Demonstration. His circular Hypotheses Mr. *Horrox* (before I could perswade him) assayed a long time with indefatigable Pains, and Study, to correct, and amend; changing and turning them every way (still amazed and amused with those lofty Titles of Perpetuity and Perfection, so impudently impos'd upon them) until we found, by comparing Observations in several places of the Orbes, that his Hypotheses would never agree with the Heavens for all times, as he confidently boasts; no, nor scarce for any one whole Year together, alter the equal Motion, Prosthaphaereses, and Excentricity howsoever you will.[25]

[24] Gascoigne to William Crabtree (25 Jan 1641), Bodleian Library, MS. Eng.7031, f.39r.

[25] *Philosophical transactions*, 1711, v.27, 288.

Both Crabtree and Gascoigne started to seek out adverse commentaries on the astronomy of Lansberge. Within less than a year of its publication in the Dutch Republic, they were scouring the highly critical *Dissertatio Astronomica* (1640) of Johannes Phocylides Holwarda. Based on his observation of the December 1638 lunar eclipse that the young English astronomers had also assiduously recorded, the 22 year old Holwarda launched an audacious assault on the lunar tables of his renowned countryman, Lansberge.[26]

At this stage Crabtree was unsure as to whether Gascoigne was convinced that Kepler provided a reliable alternative and he urged that,

> *Kepler's Elliptick* is undoubtedly the way which the Planets describe in their Motions: And if you have read his *Comment. de motu Martis*, and his *Epit. Astron. Copern.* I doubt not you will say his Theory is the most rational, demonstrative, harmonious, simple, and natural that is yet thought of, (or I suppose can be;) all those superfluous Fictions being rejected by him, which others are forced so absurdly to introduce. And although in some respects his Tables be deficient, yet being once corrected by due Observations, they hold true in the rest: Which is that argument of Truth, which *Lansberge's* and all others want.[27]

Gascoigne too was feeling let down by Lansberge. On 2 December 1640 he wrote to Crabtree promising that he would 'shortly make some jugling appeare about his [Lansberge's] diameter of the Sun, would it not bee too far a fetch'. By February 1641 he was in a position to explain to Oughtred how he had made micrometer measurements proving that the eccentricity of the Earth's orbit was more in keeping with Kepler's hypothesis than Lansberge's:

> It seems by these [solar diameter measurements] that the sun's greatest diameter is very near 33′, and, by this little, that it goes the whole eccentricity I will not be confident, until I have a greater certainty by the least diameter; although I have observations which persuade me that it keeps very near ½ eccentricity of Lansberg.
>
> The least semid. 15′53″ + 32″ = 16′25″ the greatest.
> Lansberg's least 16′47″ + 1′12″ = 17′59″ his greatest.
>
> If I had not too long trusted Lansberg and Hortensius, I should have ere this been furnished with observations sufficient to have decided that controversy between Kepler and Hortensius. At this present I am not confident of any more than this, that the greatest is about 16′25″.[28]

'Lansberge & the observations differ so very much', he remarked to Crabtree a little later, 'that I know not what to thinke'.[29] [22 Mar 1641]

[26] Dijkstra, Arjen, *A wonderful little book: The Dissertation Astronomica by Johannes Phocylides Holwarda*, in *Centres and cycles of accumulation in and around the Netherlands*, ed. Lissa Roberts (Münster, 2011), 73–99.

[27] *Philosophical transactions*, 1711, v.27, 288–289.

[28] Rigaud, S.P., *Correspondence of scientific men of the seventeenth century* (Oxford, 1841), v.1., 44. Gascoigne to William Oughtred (c. Feb 1641).

[29] The National Archives, RGO 1/40, f.14v.

It is worth noting that Gascoigne's measurements give an eccentricity, e, of 0.0165 (compared with a modern value of 0.0167). Lansberge's implied value is 0.0345—twice as much. This 'bisecting' of the eccentricity lent support to the Keplerian ellipse, as opposed to the off-centre circle of the Copernican scheme.

By means of measurements of the solar diameter, Gascoigne convinced himself of the reality of the Earth's elliptical orbit and the superiority of the Rudolphine Tables, but he still reserved judgement on Kepler's second 'law': that the imaginary line between the Earth and the Sun swept out equal areas in equal times. In fact, in the summer of 1641 [29 Jul 1641] he outlined a solar theory for Crabtree, based on a set of six 'theorems', the first of which was that 'The Center of the Earth moves equally (or equall spases Equale times in the Peripherie of the Elipsis)'. According to Flamsteed, in later correspondence Gascoigne would retract this solar theory, but no trace remains to explain why.

The opinions of Gascoigne need to be seen in the context of his constant striving for improved instrumentation. As one might imagine, his earliest observations were simply proofs that his methods were feasible. By his own admission the accuracy of the observations was not sufficient to enable any conclusions to be drawn regarding the opposing orbital theories. The fifth page of Flamsteed's *Historiae Coelestis* is devoted to a table of Gascoigne's measurements of the lunar semi-diameter throughout 1641 and part of 1642. If we compare the observed values with those predicted by modern techniques, it becomes clear that the discrepancies are initially quite substantial. During the first half of 1641 the angular semidiameter was overestimated by an average of about 46″. Still not bad bearing in mind that previous methods could probably only detect changes of a minute or two of arc. By September and October the average errors had been reduced to approximately 12″; by November, 4″; and by December 1641 the observations of Gascoigne were routinely within 2 or 3 arc-seconds of the modern value.

On 12 December 1640, barely a fortnight before his untimely death, Jeremiah Horrocks, excited by Crabtree's news of the pioneering work at Middleton, wrote from Toxteth, 'I would especially like to have some of Gascoigne's observations of the Moon's diameter, so that I might see how they agree with my new theory'.[30] Maybe they would not have been quite as useful as he imagined at that early stage.

Only by late December 1641 did Gascoigne feel able to tell Crabtree: 'Mr. Horrox his Theory of the Moon I shall shortly be furnished to try'.

Even the solar semi-diameters, though much praised by later historians and compared favourably with the *Connaissances des Temps* by Delambre, should be treated as 'work in progress'. Their reliability was not only dependent upon lens quality and the workmanship embodied in Gascoigne's instruments. It was also affected by calculating mistakes—possibly arising from the limitations of the logarithmic and trigonometrical tables available to him (as well as the constant need to use paper

[30] Horrocks, Jeremiah, *Opera Posthuma* (London, 1678), 336. Letter from William Crabtree to Gascoigne (12 December 1640). Translated in Wilson, Curtis, *Astronomy from Kepler to Newton* (London, 1989), art.VII, 78.

sparingly!). Working through the calculations for the solar semi-diameter on 31st October 1641, which Gascoigne gave in his second letter to Oughtred, it is clear that the answer should be significantly different from the 16′11″ that he gave and which was subsequently quoted by Flamsteed in the introduction to his *Historiae Coelestis*.

Part III
Digging Further

Chapter 11
The Flamsteed Papers

The papers of the first Astronomer Royal, John Flamsteed, form part of the archive of the Royal Greenwich Observatory and are now deposited at the library of the University of Cambridge, where they occupy about 7 m of shelving.[1] They are gathered into 76 pieces or batches (RGO 1/1–1/76). The first 71 are in the order imposed by Francis Baily in the catalogue given in *An Account of the Revd John Flamsteed*.[2] To these five more have been added at a later date.

In a note, bound in with the papers, Flamsteed (Fig. 11.1) relates the story of how some of the Horrocks, Crabtree and Gascoigne correspondence came to be rescued: 'after all their deaths Mr. Christopher Townley of the Car in Lancashire & Sir Jonas More went to Mr. Crabtree's house where his widow gave them all his papers. These they secured and afterwards sorted'.[3]

The Gascoigne-Crabtree correspondence transcriptions, made by Flamsteed during his 1671–1672 visits to Towneley, are largely contained in RGO 1/9 and RGO 1/40. There, in a draft treatise of more than 30 pages discussing the letters, Flamsteed includes the following prefatory remarks:

> By the singular favour of the ingenuous and right worthy Gentleman Richard Townly of Townly in Lancashire I was permitted a view of some letters betwixt that thrice Ingenuous Mr Gascoigne of Middleton in Yorkshire & the no lesse judicious Mr W Crabtree of Broughton neare Manchester. Both some while since deceased, Mr Gascoigne in our uncivill wars, & Mr Crabtree since as I hear in peace.[4]

The transcripts include extracts from, or references to, the letters shown in Table 11.1.

[1]Perkins, Adam, *The Flamsteed papers in the archives of the Royal Greenwich Observatory*, included in *Flamsteed's stars* (Ed. Frances Willmoth, 1997), 217–253.

[2]Baily, Francis, *An account of the Revd John Flamsteed* (London, 1835, reprinted 1966), lix–lxxiii.

[3]The National Archives, RGO 1/65A, f.112v.

[4]The National Archives, RGO 1/40, f.9v.

D. Sellers, *In Search of William Gascoigne: Seventeenth Century Astronomer*, Astrophysics and Space Science Library 390, DOI 10.1007/978-1-4614-4097-0_11,

Fig. 11.1 John Flamsteed (© Royal Society)

In their earliest letters—as shown by Derham's transcriptions—the two young astronomers exchange ideas about the nature of sun-spots, which were appearing relatively frequently around that time, about the transit of Venus, and about Gascoigne's extraordinary optical inventions—particularly the telescopic sight. Crabtree makes the first of many requests for written explanations of the manner in which Gascoigne's instruments work and for observational measurements—especially ones accurately fixing the moon's position at particular times.

Much of the material transcribed by Flamsteed—his selections arising from his own interests as much as those of the correspondents—shows Gascoigne and Crabtree exchanging and discussing observational measurements: above all, solar

Table 11.1 Gascoigne-Crabtree correspondence included in Flamsteed's transcriptions (the Derham references are to the Philosophical transactions)

Author	Date	RGO ref	Derham ref
Crabtree	1640 Aug 7	1/40, ff.9v-10r	v.27, 280-9
Crabtree	1640 Oct 30	1/40, ff.10r-10v	v.30, 606-8
Gascoigne	1640 Dec 2	1/9, f.4r; 1/40, ff.10v-11v	
Crabtree	1640 Dec 28	1/40, f.12v	v.30, 608
Gascoigne	1641 Jan 25	1/40, ff.12v-13v	v.30, 604
Crabtree	1641 Mar 18	1/9, f.3r; 1/40, ff.13v-14r	
Gascoigne	1641 Mar 22	1/40, f.14v	
Gascoigne	1641 Apr 6	1/40, f.14v	
Crabtree	1641 May 26	1/9, f.5v; 1/40, ff.14v-15r	
Gascoigne	1641 Jun 22	1/9, f.5v; 1/40, ff.15r-16r	
Crabtree	1641 Jul 13	1/40, f.16r	
Gascoigne	1641 Jul 29	1/40, f.15v	
Gascoigne	1641 Sep 1	1/40, f.16v	
Gascoigne	1641 Nov 1	1/40, f.16v	
Crabtree	1641 Dec 6	1/40, f.16r-16v, f.17r	v.30, 608-9
Gascoigne	1641 Dec 24	1/40, f.17r	v.30, 605
Gascoigne	'Date worn out'	??	v.30, 605
Crabtree	1642 Mar 17	1/40, f.17r	
Gascoigne	1642 Jun 8	1/40, f.20v	
Crabtree	1642 Jun 21	1/9, f.6r; 1/40, ff.17v-19r	

and lunar diameters, eclipses and positions of the moon at notable appulses. These they compare with the ephemerides and tables of Lansberg, Kepler, Phocilides, Argolus, and Hortensius. The bulk of their endeavours—at least in the transcribed correspondence—seem to be geared towards elucidating the correct orbital parameters of the earth and moon.

In a letter of 18 March 1641 Crabtree explains how he had observed the December 1638 lunar eclipse [20–21 December NS].

> The Eclipse of the Moone 10 December 1638 I observed as exactly as the dim clouds would give me leave. the time I tooke by a good brasse clock whose minute finger goes about in a circle by it selfe once every hour & being fastned upon the axis of the lowest wheele is firme without any jetting. the circle is divided into 60 minutes & every minute into 3 partes & so is discernable by 10″ of time & lesse. I had one at the clock to note the time when I gave notice & by 10 severall observations of the lunar azimuth taken by a semicircle of 4 foot diameter I found the clock to keep pace with the heavens in a constant tenor & within lesse then a minute dureing all the time of the Eclipse.
>
> ...The observations differ a little from Phocilides, but agree with yours to admiration onely the correction of the clock either in you or mee is a little missed. for the observations allow no differences of Meridians betwixt us which by itinerary distance should seeme to be about 3 min, of time. For my parte I observed both the time & Eclipse with as much exactnesse & care as I could. The time I durst say much by. yet will not be too confident in any thinge.[5]

[5] *Ibid.*, f.13v-14r. The observations of 'Phocilides' that are referenced by Crabtree are probably those published in *Dissertatio Astronomica quae Occasione ultimi Lunario anni 1638 Deliquii*, J. Phocylides Holwarda (1640).

The difference in the observing techniques of the two men is illuminating. William Crabtree (letter to Gascoigne, 26 May 1641) describes using a cross-staff, fitted with needles, to observe the sun:

> The 6 of Aprill the Sun being about 3 degrees high appeared tractable to the bare eye, through thin clouds & vapors; I fixed 2 small needles & went back exactly squarewise from the one of them till by the outside of the one & inside of the other (both wayes tried) I could just see the Suns transverse edges. I there noted the place of mine eye, which being after measured was *AB* 2011 such partes as *BD* (the distance of the needles from the out side of the one to the inside of the other) was 19.75 hence I gather the suns diameter 32–03 semid. 16-01½ The Suns limbe was not very exactly distinguishable because of its dimnesse in the cloudy vapors else I should have thought this an exact observation. for *AB* was about 5 yards distance.[6]

By contrast, Gascoigne was measuring the solar disk 'by the screws & wheeles' of the micrometer.[7]

The story of the spider's web is apt to make Gascoigne's optical discoveries appear to have arisen from pure serendipity. The transcriptions made by Flamsteed show that this is far from the truth. The previously uncited letters show Gascoigne and Crabtree discussing in great detail an analytical approach to the design of optical systems. They show Gascoigne as an earnest student of theoretical optics. He had clearly pored over Descartes' *De l'arc-en-soleil* (1637) and *De la dioptrique* (1637). He had studied critically the work of Christopher Scheiner, including *Refractiones coelestes* (1615). Possibly he had seen Scheiner's *Oculus hoc est: Fundamentum opticum* (1619) with its explanation of the anatomy of the eye. To see for himself exactly how a likeness of the external world is cast into the brain, he had performed his own experiments on an ox's eye. Descartes had described just such dissections in *De la dioptrique*—'taking the eye of a newly deceased man, or, for want of that, of an ox',[8] pointing out what could be seen 'if it were possible to cut the eye in half, without the liquids with which it is filled escaping and without any of its parts changing their places'.[9] Gascoigne's letter of 8 June 1642 to Crabtree relates how, with typical ingenuity, he found a way of avoiding these pitfalls:

> I after no few plotteings find how to performe this have at last thought upon & practised this simple yet prime way when the season serves for it. I take a fresh eye in a frosty evening & place it where it may be frozen thorow. then I can cut it as I please and each part will be very different in colour, & remaine in theire naturall site which may be pricked forth by an oiled paper. or the halfe eye being fastened to a rimmeing table so as the cut side be from the table (moveing on<->[])] whereon another rimmer beareing a glasse convex of the one or both sides with a thrid haveing a table covered with white paper are likewise moveable, then alloweing the Sun to shine through an hole of 3, 4 or 5 inch breadth into a roome otherwise darkened turneing the cut side of the ey to the Sun passeing by the other 2, you may by removeing the table covered with paper find the true place of representation, which will if

[6] *Ibid.*, f.14v-15r.

[7] *Ibid.*, f.17r.

[8] Descartes, René, *Discourse on method, Optics, Geometry, and Meteorology*, translated by Paul J.Olscamp (Indianapolis, 2001), 91.

[9] *Ibid.*, 84.

your glasse be great & good appeare not discerneably infferior in appearance to its object, & for magnitude thus. If the distance betweene the ½ eye and glasse exceed that of the glasse & table the representation will be lesse for extent but brighter for colour. If otherwise the contrary. When you have this as great as you please you may by your pen draw each part upon the paper. Or at any time I can describe the prime parts thus having a glasse & table fitted to observe the Sun's spots I place an eye with the horny tunicle either upward or downward betweene the inmost glasse & the table. So neare the glasse as your eye will allmost fill up the compasse of the Sun's image. Then the representation of the eye will be large (proportionable to the Sun's image) upon the table and thus I pricked out these 3 figures of the horny tunicle Christalline outward & inward humor, which is for ought I yet know wholly unknowne to the world ...[10]

Gascoigne goes on to record the measurements that he has taken of the various parts of the eye and to describe their role in refracting the incoming rays. These, however, are not transcribed: 'the scheme [diagram] of the eye is wanting therefore I omit those measures of it which hee tooke, by observeing it in his long tube, with his screw', says Flamsteed, who cannot help adding that this experiment shows 'the sublimity of [Gascoigne's] wit & profoundnesse of judgement'.

Henry Power must have been shown this letter by his friend Richard Towneley some time previously, for he includes in his *Experimental Philosophy* (published 7 years before Flamsteed made a copy) a description of the procedure with the frozen eye, stating that he obtained it from 'some fragments and papers' of Gascoigne. It was, he declared admiringly, 'one of the ingenious excogitations of M.Gascoign's'.[11]

Disappointingly, Flamsteed's transcriptions generally exclude such 'small talk' and personal information as might have been part of the original letters, leaving one to guess at the relationships and circumstances that would have been revealed by such content: Flamsteed was no Aubrey. With typical brevity, after his transcription of parts of Crabtree's letter of 18 March 1641, Flamsteed adds:

I must not omit that in the very preface of this letter hee spends one whole page in lamenting the death of Mr Horrox whom thence I guess to have died in February precedeing ...[12]

Elsewhere in his papers, Flamsteed gives some very brief biographical details, including a different date for Horrocks' death, saying that he 'died Jan 3 1641 I have heard say of an Apoplexy aged 23 years'.[13] In a short note touching on information provided by Christopher Towneley he says of Gascoigne:

...hee had something a sowre countenance but was much a gentleman in his behaviour. hes father was much given to mechanick by which he hoped to performe no meane things & once promised to show all sorts of manifactures workeing by them. had a large house filld full of his tooles & contrivances. of which if any were removed out of their places hee would readily misse them. Such was his memory: his Estate was 8 or 900 ld p Annum.

[10]The National Archives, RGO 1/40, f.20v.

[11]Power, Henry, *Experimental Philosophy* (London, 1664), pp.80–1.

[12]The National Archives, RGO 1/40, f.14r.

[13]The National Archives, RGO 1/65, f.112v.

Christopher Towneley had written to his nephew, Richard, that

> I have done all that lies in my power concerneing Crabtree. but I cannot find what I conceive you desire but this. William Crabtree lived at Broughton neare Manchester, hee bought cloth and baises, and sent them to Blackwell Hall,[14] thus much for his trade. hee died as neare as I can conjecture about 1643 but I can not exactly tell the moneth. hee left a wife & one sonne which are both dead.[15]

The treatise that Flamsteed intended to produce, based on the Gascoigne-Crabtree correspondence, came to nought. It remained as an unfinished draft: Incomplete, but the repository nevertheless of the best surviving overview of the 3 or 4 year interchange between the two young astronomers.

Sitting amongst Flamsteed's papers from the Royal Greenwich Observatory, however, is another document bearing substantial witness to the discoveries of Gascoigne and their impact on the first Astronomer Royal. It was described by Flamsteed's biographer, Francis Baily, simply as 'Flamsteed's Lectures on Astronomy read at Gresham College in 1681–1684. Small thick Quarto, bound in a marbled cover'.[16] As to the contents of the lectures (39 in total), Baily provided no elaboration. His main biographical effort was focused on rescuing Flamsteed's posthumous reputation from calumnies, as he saw them, perpetrated by Isaac Newton and others, who had criticised the extreme delays in the publication of the observatory's data. The even-handed role of Baily's book, in his own mind, was to ensure that

> … whilst on the one hand the brilliant talents of Halley and the still more transcendent mind of Newton have been set in opposition to the less splendid yet very useful labours of Flamsteed, and arguments drawn from this relative position of the respective parties, we see that on the other side the ardent zeal of the Astronomer Royal, his piety, his accuracy, his integrity, and his independent spirit, have been arrayed against the alleged misconduct of his two powerful opponents.[17]

Perhaps, because of the overwhelming reputation and achievements of Newton, the life of Flamsteed and the contents of his archives have stayed relatively ignored and unexplored until recent times.

---

[14] Blackwell Hall was situated at 6 Basinghall Street, London, adjacent to the Guildhall, and was a market place for woollen cloths. It was destroyed in the Great Fire of London. Though rebuilt, it was removed in 1820 to make way for new law courts at the Guildhall.

[15] The National Archives, RGO 1/40, f.22r. The year of Crabtree's death and the number of surviving children vary in other accounts. See, Crofton, Henry T., *Broughton near Manchester. Its Topography and manor court*, Chetham Miscellanies, New Series, ii (1909), 37–41. Here, 1 August 1644 is put as the date of Crabtree's burial and he is said to have had one son and three daughters. Whether the William Crabtree buried on that date in the precincts of the Manchester Collegiate Church was indeed the astronomer, however, is not beyond question. The date of the probate for the '*Will of Wm Crabtree, Clothier of Broughton, Lancashire*' at the Prerogative Court of Cantebury is given as 22 Sep 1645 (The National Archives, PROB 11/194).

[16] Baily, Francis, *Op.Cit.*, lxviii.

[17] *Ibid.*, 675.

The Gresham Lectures remained unpublished until 1975, the 400th anniversary year of the Greenwich Observatory, when they appeared in a volume edited by Professor Eric Forbes (1933–1984) of the University of Edinburgh.[18]

The prescription for the astronomy lectures at Gresham College laid down by a committee in 1597 was still the one in force as Flamsteed started to plan what he would cover. It decreed that the lecturer should concentrate on traditional matters: the principles of the sphere, the motions of the planets and the use of navigational instruments. But a revolution in astronomy had taken place during the intervening years, in which the telescope, the micrometer eyepiece and other optical devices had played a pivotal role. Flamsteed was determined to do justice to the new astronomy and his lectures enable one to glimpse the progress of astronomy to his time and to see the course of his own intellectual development as part of that.

In particular, Flamsteed used some of the earlier lectures to bring Gascoigne's optical ideas to greater notice, including the use of the sine law of refraction to determine the path of light rays through plano-convex and plano-concave lenses. The law itself was still absent from many of the texts that students were relying upon. Flamsteed regarded this as a serious weakness and in the fourth lecture, commenting on the *Dioptricks* of Honoré Fabri (1607–1688), he warned

> let me advise those that peruse this acute author to be very carefull how they admit any thinge on his authority for hee supposes the refraction in glasse to be 1/3 of the angle of incidence whereas Mr Gascoignes experiments Mr Newtons & those of our Geometry Lecturer [Hooke] all prove that Not the angles but their sines are proportionall[19]

This *application* of the law was the key to any progress at all, as he affirmed to Molyneux in a letter midway through the lecture series. Here he confided that he had 'learnt his optics'

> from some loose papers of one Mr Gascoigne … in lesse than 2 hours time and all that I know of glasses is but a superstructure on that foundation Nor doe I doubt but if you will please to take some little paines in prosecuteing the first principles you will find suitable successes. before I met with these I was, as you, tormented with readeing the discourses of Opticall writers, who I now find understood not the prime proposition and therefore onely gropeing in the Darke themselves could never lead others aright. They were ignorant of that proportion Which the Angles of incidence beare to the Angles refracted or if they knew some little of it they perceived not how it might be applyed to glasses.[20]

Above all, the progress that Flamsteed had in mind was the use of telescopic sights on astronomical instruments and the use of the micrometer to address some of astronomy's biggest questions. In the sixth lecture, referring to Gascoigne, he

[18] Forbes, Eric G., *The Gresham Lectures of John Flamsteed* (London, 1975).

[19] *Ibid.*, 127.

[20] Forbes, E.G., Murdin, L., and Willmoth, F., (Eds), *The correspondence of John Flamsteed, The first Astronomer Royal*, v.1 (1666–1682) (Bristol & Philadelphia, 1995), Letter to Molyneux, 29 May 1682, 897–8.

related how a start was made with the telescopic sight (and then came to an abrupt end):

> ... something more than 40 years since. A ingenuous English gentleman examineing the Nature of opti[c]k glasses more diligently found that if instead of the Concave eye glasse hitherto used hee should apply a Convex eye glasse; the telescope would not onely receave a greater Angle with the heavens but also that hee could clearely perceave when the object passed over a needle thrid or point placed into common Focus & that therefore if such a telescope were placed on the index of an instrument the diameters & distances of the planets might be thereby exactly measured. his First attempts were to take the Diameter of the Luminaryes & planets in a single tube. his successe therein encouraged him to proceed further; at length hee caused a sextant to be made & then began to employ it when our late unnaturall war commenced in which hee became a party & was slaiein ...[21]

In the ninth lecture, the audience was shown a practical example of the giant leap that had been made possible by extending the same principle to create a micrometer eyepiece for attachment to the keplerian the telescope. Flamsteed described the previous techniques that had been used to measure the seasonal variations of the angular diameter of the Sun. Using the naked eye, astronomers had had to resort to timing (with a pendulum) the passage of the Sun across a marker, straight edge, or wire, held in front of the observer. The best of the techniques had been perfected by Gabriel de Mouton, 'an ingenuous Frenchman', who had projected a magnified image of the Sun onto a screen and counted the vibrations of a pendulum whilst the image drifted across a line drawn perpendicular to the direction of drift. Through frequent repetition of the experiment, de Mouton had obtained apogee and perigee diameters of 31′45″ and 32′32″.

> But the best method, *declared Flamsteed*, was that of Mr Gascoigne who tooke his measures by the helpe of a Micrometer or payre of screws placed in the inner focus of his Telescope whereby hee found the Diameter of the Apoge sun 31′40″ of the Perige 32′50″ with which Mr T[owneley']s agree. with a telescope of about 14 foot & something better contrived mycrometer at Derby in the Yeare 1672 I determined the breadth of the Apoge sun 31′30″ or some little more in his Meane distance on the 19th of March 32′12″ in his Perige 32′40″ with which Mr Cassini assures mee the French observations agree exactly. whence it will follow that the eccentricity of the Sun is bisected ...[22]

The Flamsteed correspondence[23] that has been published between 1995 and 2002 has thrown up many additional references to Gascoigne (some of which have been already quoted in the present work), but the Flamsteed papers in the Royal Greenwich Observatory archives still hold much relevant material that is yet to be mined, including notes by Sir Jonas Moore relating to Gascoigne and Crabtree.

---

[21] Forbes, Eric G., *The Gresham Lectures of John Flamsteed* (London, 1975), 148–9.

[22] *Ibid.*, 197.

[23] Eric G. Forbes, Lesley Murdin and Frances Willmoth, *The Correspondence of John Flamsteed, the first Astronomer Royal*, 3 vols, covering 1666–1682, 1682–1703 and 1703–1719 respectively (Bristol and Philadelphia, 1995–2002).

Chapter 12
The Towneley Papers

Since Wheater published in 1863 his plea for help in uncovering more information about Gascoigne, a number of developments have made it potentially much easier to pursue such a goal.

Over the intervening century and a half many manuscript collections have emerged from private obscurity into public hands—or at least have temporarily surfaced in sale rooms. Large numbers of such collections have been catalogued and indexed. Digital catalogues of the National Archives, the British Library, the Bodleian Library and other important archives are now searchable via the Internet. Many antique books are available in print-on-demand or digital versions. Modern researchers can dispense with the laborious letter writing to chief librarians that was a source of so much vexation and misplaced trust for earlier inquirers. When Sidney Gaythorpe wrote to the City Librarian of Leeds in 1919, asking whether there was any mention of Gascoigne in the library copy of Thoresby's *Ducatus Leodiensis*, the emphatic reply stated that no trace of a mention could be found[1]—despite the fact that the volume contained a full page pedigree of the Gascoigne family!

A wealth of historical archive material has been assembled by new institutions—such as the Borthwick Institute in York—which have been busy preserving invaluable manuscript material, including previously inaccessible legal records.

Since 1863 much material that previously existed only in manuscript form has been transcribed and published—thus allowing its dissemination to many centres of study. For example, seventeenth century visitation records, along with their unique pedigree listings, and parish records of births, deaths and marriages, are now available in good libraries. Correspondence and diaries of Gascoigne's contemporaries and immediate succeeding generations have been published. More and more surviving documents from the civil war period have been discovered.

How does all this new material help in the search for William Gascoigne?

[1]University of Illinois, History Department, list of the *Sidney B. Gaythorpe Papers, 1903–1944, 1954, 1961* (Record Series No. 15/13/50), letter of 30 August 1919 from Leeds City Librarian.

D. Sellers, *In Search of William Gascoigne: Seventeenth Century Astronomer*, Astrophysics and Space Science Library 390, DOI 10.1007/978-1-4614-4097-0_12,

Table 12.1 Gascoigne-Crabtree correspondence included in the Bodleian transcriptions

Author	Date	Bodleian ref
Gascoigne	1640 Dec 2[a]	MS.Eng.7031, ff.43r-44v
Gascoigne	1641 Jan 25[a]	MS.Eng.7031, ff.35r-39r
Crabtree	1641 Mar 18	MS.Eng.7031, ff.29r-32r
Crabtree	1641 Mar 26	MS.Eng.7031, ff.32v-33r
Crabtree	1641 Jul 13	MS.Eng.7031, f.33r
Crabtree	1644 [sic] Dec 22	MS.Eng.7031, f.29r
??	Undated	MS.Eng.7031, f.41r

[a] Undated, but containing common material with Flamsteed transcripts of these dates.

Amongst the Towneley papers in the Bodleian Library a significant unexplored repository of copies of the letters passing between Crabtree and Gascoigne is to be found.[2] The papers were bought by the library at an auction held by Sotheby's in July 1985 (lot 546), with the aid of grants from the V & A Purchase Grant Fund, the Friends of the National Libraries and the Royal Society. Although the particular papers of interest here are catalogued simply as 'copies of extracts from letters from Mr. Crabtree, 1640–1644, n.d., relating to astronomy', it is clear from the content that there are six letters, comprising 23 pages, which are actually copies of four letters from Crabtree to Gascoigne and two from Gascoigne to Crabtree (Table 12.1).

In the first of these letters (2 December 1640) Gascoigne explains, with example calculations, the meaning of the five diagrams that he had previously sent to Crabtree. The explanation (without the calculations) had also been included in the Flamsteed transcriptions. The calculations, similar to those summarised in the second letter to Oughtred, show how the apparent size of the solar disk may be determined both by projection and directly by means of the micrometer: a 'moveable brasse scale observed by a greene glasse interposed betweene the eye … and the convex glasse [of the eyepiece]'.[3] Gascoigne is at pains to point out that he 'dare not affirme these observations very true, onely here intending to declare how wee may make true ones'.[4] He is clearly still in the process of perfecting his instruments and the Bodleian transcript (but not Flamsteed's) includes his belief that 'Mr. Towneley thinkes he can procure me a peice of Iron to my mind to forme glasses on as I think best, which if I get I doubt not in a short time to be furnished with glasses of speciall power'.[5]

The letter from Crabtree dated 22 December 1644 is surely misdated (by that time Gascoigne was dead). The content marks it as one of the earliest between the

[2] Bodleian Library, *MS. Eng. c.7031*, ff.29–49. Catalogued as 'astronomy papers, mainly of Christopher Towneley' within which ff.29–44 are catalogued as 'copies of extracts from letters from Mr Crabtree, 1640–1644, n.d., relating to astronomy, [c.1661 – see inscription on fol.33]'. At the foot of f.33r is a lightly written pencil note in a more modern hand 'The Water Paper Mark 1661. Archaeologia Vol:12'.

[3] *Ibid.*, f.43v.

[4] *Ibid.*, f.43r.

[5] *Ibid.*, f.43r.

two: Crabtree is still seeking further explanation of the diagrams and promises 'soe shall I always acknowledge you as my sole instructor in these optick secrets'[6].

From Gascoigne's letter of 25 January 1641, Flamsteed had chosen to transcribe only the prime observational data and a table of angles of refraction (less than 60 lines). The Bodleian transcript of this date, however, runs to nine pages. It includes a discussion of the ideas of Kepler and Descartes on refraction, a lengthy explanation of Gascoigne's experimental procedures, and comparisons of his observations with the theories of Lansberg, Kepler and others.

He explains how he measures the radius of curvature of his convex lenses by adjusting the curvature of a fine pasteboard cut-out until the cut-out neatly fits against his lens—'which labour', he adds, 'were needlesse if we know the axis of the toole forming the glass'.[7]

As to his own tool, it has become clear that Towneley will not be able to deliver the required 'peice of Iron' and a substitute may depend on the ravages of alcohol!

> I had a letter from Mr Towneley who cannot meet with Rifler the best glasse grinder, nor a man that can forge iron into any usefull toole for my way, which must have it void of flawes, I have given directions to a Trooper an excellent Gunsmith whose skill as soone as Ale will allow I shall see. It may be ere long I shall have have occasion to visit London where I shall fit my selfe. If you come at any time neare Wigan inform your self as fully as may be how many pounds of brasse they can melt into any forme if they could melt brasse to cleave to iron as I have seene some great ordnance it would be much cheaper to have a quadrant cast at an iron furnace & after on one side covered with brasse, on the rimme thick enough to receiv a mult rimmed cursor, which you will find readily affording 10 or 100 more divisions then those 2 in Morinus[8]

Although Gascoigne does not give the optical details of his keplerian telescope, these can perhaps be inferred from the advice that he gives to Crabtree, who is hoping to get Towneley to procure suitable lenses to create a similar one. The eyepiece biconvex glass, he says, should be about two inches diameter and the thickness 0.3 in.. The radii of curvature of the surface nearest the objective should be 19.00 in. and the opposite surface 3.46 in.. Assuming a refractive index of 1.574—in common with his planoconvex objective—the focal length would be about 5.1 in.. His observational results imply that the objective would have had a radius of curvature of 25.8 in. and a focal length of 45 in.. The diameter of the objective is not stated.[9]

Gascoigne refers frequently to the work of ancient astronomers as well as contemporary writers such as Descartes, Kepler, Lansberge, Hortensius, Scheiner, Malapert, and Morinus. He obviously has a substantial library of his own, but this

---

[6] *Ibid.*, f.29r.

[7] *Ibid.*, f.36r.

[8] *Ibid.*, ff.37r-37v.

[9] *Ibid.*, f 37r. Sidney Gaythorpe was able to deduce the lens parameters for Gascoigne's galilean telescope (see chapter 9). However, based on the letters to Oughtred – the only available letters at the time - he had to concede that "notwithstanding its great historic interest and importance as being the first Keplerian telescope to which a micrometer was was fitted, there is, unfortunately, an almost entire lack of information as to the optical details of this instrument".

letter shows how he is making use of loans from Crabtree as well as books supplied from London.

> I have returned your Keplers, but keepe your Sirturus[10] for a review when my Iron is finished, & let me once for all assure you that what so ever I doe or can master shall be at your command. If you please to get me those 2 bookes Suelly obser: Hassi: Kepleri harmon mundi (because I named not this amongst those of his I sent for to London) if it have any thing in it but meerely speculation, I shall returne what you disburse for them by him I sent my last letter to you by.[11]

The substantial letter from Crabtree of 18 March 1641, from which Flamsteed had copied only an account of the lunar eclipse of 10 December 1638, is in this version almost exclusively about optical theory. It commences with Crabtree's doubts about the usefulness of relying on the analogy of a balance to represent refraction, despite Kepler's words—'I love analogies most of all: they are my most faithful teachers, aware of all the hidden secrets of nature'—which Gascoigne had quoted approvingly in his 25 January letter.[12] Nevertheless, he acknowledges that Gascoigne's measurements of the passage of light from air to water clearly demonstrated that the 'refractions there are exactly determined according to the sines of Angles'.[13]

Crabtree continues his plea for Gascoigne's patience, as he grapples with new optical concepts (particularly the reasons why only certain rays are selected for inclusion in the ray dagrams).

> You have taken a great deale of paines for mee in explaning these Demonstrations to drive them into my unprepared conceit, I was a meere stranger to the very first Notions of these things before I received them from you …
>
> I beseech you therefore as you have begun with mee, take so much paines more as to make me understand mine error herein, as plainely as you can for those things which by reason of your versednesse in this way are common & easy to you, are by reason of my strangenesse therein hard & obscure to me. If there be any thing in your former letters which will readie me, if you but explaine it & refer mee to it, I shall spare no diligence in comparing, For though I am not like to get requisites to make use of those things, or but few of them, yet since you have accepted mee for your scholler I would faine learne my lesson perfectly. Neither shall I be ashamed to show my ignorance freely …[14]

'I pray you pluck mee by the eare where I am amisse',[15] he requests in a later letter.

---

[10] Probably *De telescopio*, Hieronymus Sirturus (Frankfurt, 1618).

[11] Bodleian Library, *MS. Eng. c.7031*, f.39r.

[12] *Ibid.*, f.35r. Gascoigne quotes the Latin text "plurimum namque amo analogias fidelissimos meos magistros, omnium naturas arcanorum conscios" from Kepler's *Ad Vitellionem paralipomena* (Frankfurt, 1604), 95. The English translation is from *Kepler's Optics*, William H.Donahue, (Santa Fe, 2000), 109.

[13] *Ibid.*, f.29v.

[14] *Ibid.*, f.29v-30r (Crabtree to Gascoigne, 18 March 1641).

[15] *Ibid.*, f.32v (Crabtree to Gascoigne, 26 May 1641).

The suggestion by Gascoigne that a glass grinding instrument could be forged does not seem ideal to Crabtree, though—if that be the only way—he offers to use the services of a gunsmith near Broughton.

> I thought your instrument for forming glasses should have beene cast of iron or at least made of Cast Iron. If forging will doe it, we have a Gunsmith some 3 miles from mee (with whom I have some acquaintance & could have him doe any thing for mee) who for complait & solid workmanship & especially in that everie point of forging is thought (by many judicious workmen) to be second to none in England, if his best endeavors may any way stead you assure your selfe I will take course you shall not be awanting in anything you desire. About a fortnight hence I shall goe to Wigan & then I shall what they can doe about casting brasse.[16]

Whilst Crabtree willing confesses his ignorance on the practical application of ray theory, he is nevertheless well versed in general theoretical matters and these letters show how much he is providing valuable information to Gascoigne on the latest continental works in the field. From the writings of Claude Mydorge, he is aware of spherical aberration—the impossibility of achieving a common point of focus for all the parallel rays striking a lens with a spherical surface:

> It seemes the superficies of the glass *ab*, ought rather to be composed of segments of Ovals than of circles, Mydorgius the Frenchman hath lately written a booke in latine concerning Conicall sections which he esteems a fundamentall one to Catoptrickes & dioptrickes, but applyes it not, onely promiseth to doe it hereafter in other bookes to come forth. I have sent for the booke.

The same point had been emphasised by Descartes (see Chap. 9), whose works Gascoigne had studied intensely. There was nothing that could be done about this for the time being. Even 40 years later, Flamsteed was convinced that the solution—elliptical and hyperbolic lens manufacture—was 'only speculative & not likely to be ever brought to practice'.[17]

The real strength of Crabtree is in predictive astronomy and Gascoigne is increasingly relying upon this for his observing programme.

> As soone as I have dispatched my intentions for the solar theory I shall review them & send you some other observations, because your Ephemerides is soe ready if you will set me downe some of the Moon's passages as you did heretofore especially those that will be at evening before 10^h I shall be as observant of them as I possible can. I expect every day a case which I will order the best that I can for exquisite observation. If there be any other notable conjunctions I will rely upon you for information.[18]

One of the letters is undated and its is not clear whether it is to Crabtree from Gascoigne or vice-versa, but it does show how the young astronomers were scouring Latin accounts of ancient observations in order to scrutinise effectively the tables of Lansberge and others.

[16] *Ibid.*, f.31v (Crabtree to Gascoigne, 18 March 1641).

[17] Forbes, E.G., *The Gresham Lectures of John Flamsteed* (London, 1975), Lecture 7, 2 November 1681, 162.

[18] Bodleian Library, *MS. Eng. c.7031*, f.39r (Gascoigne to Crabtree, 25 January 1641).

> Most estimated friend, In my last letter I promised in my next to answer to divers parts of yours in that unmentioned. First I can more easily increase then cleare your doubts about the anciently observed equinoxes. Scaliger as most others prefers Hypparcus observations before Ptolemies...But our Countryman Lidiat Scaligers firce Antagonist vilifies Hipparchus...[19]

The correspondent gives extended quotes in Latin from works by Joseph Scaliger (1540–1609) and Thomas Lydiat (1572–1646). Delving into the disputatious topic of ancient chronology, it is no wonder that his confusion was increased. Controversies were fierce in the extreme, with the Oxfordshire clergyman and mathematician, Lydiat, being reviled in Scaliger's Epistles, according to the historian Henry Hallam, 'as the most stupid and ignorant of the human race, a portentous birth of England, or at best an ass and a beetle, whom it is below the dignity of the author to answer'.

The next letter from Crabtree to Gascoigne, continuing the discussion about the latter's optical diagrams, was written on 26 May 1641. It bears a light pencil note (presumably made by staff at the Bodleian) explaining that the water mark shows that the paper dates from 1661. Thus, the hand-written transcriptions were made at least 20 years after the original letters were penned.

The journey towards mutual understanding of the new optical theories was protracted. By mid-July 1641 Crabtree is feeling happier about some of Gascoigne's ideas, but, for those that he still doesn't apprehend, his own primitive lenses mean that he is at a loss for any means of putting them to the test: 'Neither can I hope with my irregular glasses to try any thing in the Sun to helpe me to any certainty herein'. He will probably have to pay a visit to Middleton to get to the bottom of matters: '... if [I] perceive not your meaning perfectly at a further review of your lines, I shall hope to be fully satisfied by ocular demonstration when I see you ...'[20] Whether such a visit actually took place, and indeed whether regular meetings played any part in their collaboration, is not recorded.

[19] *Ibid.*, f.41r.

[20] *Ibid.*, f.33r (Crabtree to Gascoigne, 13 July 1641).

Chapter 13
British Library Manuscripts

The British Library (until 1973 part of the British Museum) holds more than 310,000 volumes of manuscripts and, just as at the Bodleian, digital cataloguing means that simple 'on-line' searches are possible. Here too, previously overlooked Gascoigne material can be quickly located: One piece has been there since the very inception of the Library, having previously belonged to Sir Hans Sloane (1660–1753), one time President of the Royal Society.

By the time of his death, Sloane (Fig. 13.1), an eminent physician and naturalist, had amassed a truly outstanding collection of manuscripts and artefacts. His manor house in Chelsea, according to his friend William Stukeley (1687–1765), was 'full throughout: every closet and chimney, with books, raritys, etc.'[1]

His house became a de facto museum—as he readily welcomed a constant stream of visitors all eager to see its exhibitions. His zoological collection alone contained more than 21,000 specimens. In his library he had gathered nearly 50,000 books and 4,100 volumes of manuscripts.

In his will Sloane had expressed an earnest desire that, after his decease, his magnificent collection 'tending many ways to the manifestation of the glory of God, the confutation of atheism and its consequences, the use and improvement of physic, and other arts and sciences, and benefit of mankind, may remain together and not be separated, and that chiefly in and about the city of London … where they may by the great confluence of people be of most use'.[2] In order to give effect to this desire, he proposed that his collection, which he conservatively valued at more than £50,000, be sold to the King for the sum £20,000. The proceeds would go to his daughters.

The response of King George II was that he doubted that there was sufficient money in the Exchequer for such an enterprise. Sloane's trustees then appealed directly to the British Parliament, where many members saw this as an ideal

[1] Stukeley, W., Gale, R.& S., etc, *The Family Memoirs of the Rev. William Stukeley, M.D.: and the antiquarian and other correspondence of William Stukeley* (Durham, 1882), i, 358.

[2] *The Will of Sir Hans Sloane, Bart.*, (London, 1753), 3.

D. Sellers, *In Search of William Gascoigne: Seventeenth Century Astronomer*, Astrophysics and Space Science Library 390, DOI 10.1007/978-1-4614-4097-0_13,

Fig. 13.1 Sir Hans Sloane statue in the Chelsea Physic Garden (Drawing from *Early Science in Oxford*, R.T. Gunther, xi (1937))

opportunity to set up a national museum. Consequently, within 6 months of Sloane's death the *British Museum Act* was passed.[3] This allowed purchase of the collection and a building (Montagu House in Bloomsbury, London) to house it and a number of other publicly owned collections.

Sloane had acquired a fragmentary copy of a letter (or several letters) from William Crabtree to Gascoigne (at least part of which is dated 1 November 1641).[4] A letter of this date is referred to briefly by Flamsteed in his transcriptions, but only insofar as it contains a description of the micrometer screws and a few specific observations. The Sloane copy, though itself incomplete, extends to 15 pages. Unfortunately, it is in parts illegible—a point which is referred to in an interesting postscript written in a different hand.

> These are some of Mr William Crabtrees letters (who lived at Broughton neare Manchester in Lancashire[)], sent by him to our famous Mr Gascoigne of Midleton neare Leeds in Yorkshire: but they are so rudely and illiterately coppyed out that it is very difficult to understand the meaning of their first Author by them.
>
> Written An: Dom: 1664 Henrie Powe[r] M^{a}.D^{r} [5]

[3] *The British Museum Act* (26 Geo.II, ch.23) was passed on 7 June 1753 and Montagu House was bought in 1755.

[4] British Library MS, Add. 4021, Sloane Collection, ff.54–62, Letter from William Crabtree to William Gascoigne, 1 November 1641 (copies).

[5] *Ibid.*, f.58v.

The signatory to this comment is Henry Power: the same individual who 2 or 3 years later persuaded Richard Towneley to contact the Royal Society in response to Auzout's claims. The handwriting is identical to other extant samples of Power's.[6]

'Rudely' copied out though they are, it is possible to see from the pages of the Sloane manuscript that the ideas exchanged between Gascoigne and Crabtree extend far beyond the instrumental and observational matters that Flamsteed chose to preserve. Above all, the discussion centres on the precise character of the terrestrial orbit: the length of the tropical and equinoctial years, the motion of the apogee, the bisection of the eccentricity, the comparison of the classical theories of Hipparchus and Ptolemy with those of Tycho and, particularly, Kepler.

Although Gascoigne's precise measurements of the apparent solar diameter are given as evidence for Kepler's elliptical orbits, it is clear that his second and third laws are not yet understood or embraced. Gascoigne was striving to develop his own theory of the Sun (or the earth's orbit) and, in common with others, viewed Keplerian theory as something which still required observational verification. Crabtree gives a summary of his understanding of the Yorkshire astronomer's theory

> The principles of your Theory as I understand it … are comprised in the six Theorems
>
> 1. The Center of the Earth moves equally (or equall spases Equale times in the Peripherie of the Elipsis …)
> 2. The axis of the earth keepes alwaies paralell to it selfe (or is pararell) in one parte of the Orbe in what it was in another.
> 3. The axis of the earth hath ever the same inclination to the Axis of the Cone or to A line paralell to the Axis of the Cone.
> 4. The yeare of the Orbe is the time in which the earth moves in the peripherie of the Eleipsis from Apogeum to Apogeum, or from any other point in the peripherie of Ellipsis to the same point againe.
> 5. The Equinoctiall yeare is the time in which the earth moves in the peripherie of the Ellipsis from vernal equinox to vernal equinox. which is so much short of a whole Revolution as the motion of the Apogium in the meane time hath drawne away that point of the Eclipsis which was in the Equinoctial the yeare before from the poynt where it then was.
> 6. The arches with the apogeum moves in the Suns Orbe in that time which the earth makes a revolution from vernal equinox to vernal equinox are alwaies equall one to another, & this I call the Annuall Motion of the Apogeum: Hence will Folow the parradox that though the Apogeum moves alwaies Equall Arches in the Suns Orbe in the time of the earth's revolution from vernal equinox to vernal equinox yet it moves not altogether equall arches in Equall time because of the time of the earths revolution from vernal equinox to vernal equinox is not alwaies equall which appeares thus.
>
> The same constant Arch of the Apogeums Annuall motion in the Suns Orbe, Answers to a greater Arch of the Ellipsis neere the Apogeum then neare the perigium (as you may see if you looke upon your Cone now the arch of the Elipsis beinge time[)], The greater arch must needs bee longer time then the lesser Arch, So the difference betweene the equinoctiall

[6] Good examples of Power's handwriting are shown in Figs. 4 and 5 in *Observations and experiments of Dr Henry Power (1626–1668) supporting William Harvey's De motu cordis*, J.T. Hughes, Transactions of the Halifax Antiquarian Society, new series xiii (Halifax, 2005), 34–46.

yeare & the yeare of the orbe is sometimes more sometimes lesse, but the yeare of the Orbe is alwaies Equall therefore the Equinoctiall yeare (or the yeare/time in which the earth goes from vernal equinox to vernal equinox of the Sunes Orbe) is unequall in that respect & was shortest in the creation wen the vernal Equinox fell about the Apogeum, and now growes lesse as the Equinoxe drawes nearer to the perigell.[7]

Crabtree found the theorems persuasive to a point:

Thus far I went smothly on with the theory and was well satisfied in all thinges But as I was writinge these last lines before goinge & thinkinge againe att it My meditations knitt me such a gordian knott as I had much adoe to unloose, for I perrseved that upon the six former Theorems there would nessesarily follow A strange inequalitie in the Apparent equinoxes & solstises, which would carrie them with a kind of Liberation from their equall plases in the Suns Orbe, sometymes one way sometymes Another.[8]

The solar theory that Crabtree is struggling with must be that which Gascoigne had proposed in a letter (29 July 1641) 3 months earlier. Flamsteed had transcribed the letter, but omitted the section dealing with this matter, noting only,

after this hee proposeth a solar theory & (holdeing the obliquity of the Ecliptick mutable) an hypothesis to salve it. But these I find in one of his last letters hee retracted it, therefore I pretermit it.

Crabtree's letter was typically wide-ranging, and this was at least in part due to the irregular nature of the postal system. It was essential that each letter kept discussion going on multiple fronts. A purely sequential discussion, dealing with one issue at a time, would take forever. Crabtree advised his friend,

I am forced to conclude else I loose this weeke also I shall exceedingly desire to heare from you. You may direct the letters before either to Richard Martincrofts in Manchester, or to the Clerks house in Rachdale. Those that come through Manchester are some tymes a good while before they come to mee.[9]

In terminating another letter in this manuscript, Crabtree incidentally reveals something of Gascoigne's state of health: 'Could all of this my kinde freinde applied unto your legge restore it perfectly sound you should not God willinge longe bee lame …'[10]

Since the time of the founding collection, the British Library continued to receive—through benefactors, bequests and death duties, as well as purchase—the contents of many lesser collections that were formerly in private hands.

Thus in 1980 the Library acquired some papers that had once belonged to Frederick North, 2nd Earl of Guilford (1732–1792): the British Prime Minister from 1770 to 1782—reputedly the Prime Minister who 'lost America'. In one bundle of these papers—now item 61,873 in the Library's additional manuscript collection—

[7] British Library MS, Add. 4021, Sloane Collection. f.56v.

[8] *Ibid.*, ff.56v-57r.

[9] *Ibid.*, f.58r.

[10] *Ibid.*, f.58v.

Fig. 13.2 Sir Kenelm Digby (Author's collection)

is a letter in William Gascoigne's own handwriting, dated 30 November 16[40] and addressed to Sir Kenelm Digby (Fig. 13.2).[11]

This letter to Digby is interesting as it shows Gascoigne corresponding with wider circles and not only explaining the geometrical basis for his optical schemes, but also making clear the relevance of his micrometer to the key astronomical questions of the day.

Sir Kenelm Digby (1603–1665), according to Aubrey's *Brief lives*, 'was held to be the most accomplished cavalier of his time'. He was an early member of the Royal Society, was said to understand ten or more languages, and wrote on a wide variety of scientific, philosophical and religious topics. By the 1630s his fame was running high, but his position as a Catholic intimate of the King and Queen incurred great suspicion in some quarters—not least, because his father, Sir Everard Digby, had been one of the 'Gunpowder Plot' conspirators. Indeed, Digby senior had been

[11] British Library MS, Add. 61873, North (Sheffield Park) Papers, xiv, ff.54–55, Letter from William Gascoigne to Sir Kenelm Digby, 30 November [1640?]. This was amongst the papers purchased of Mrs K.F. King of Fletching by the Museum on 12 September 1980. A transcription and assessment of this letter is available in *A letter from William Gascoigne to Sir Kenelm Digby*, David Sellers, Journal for the History of Astronomy, xxxvii (2006), 405–16.

the first of the conspirators to be executed. He was dragged behind a cart to the place of execution, outside St Paul's Cathedral, whilst his wife and infant sons looked on. Then, as he was cut down from the scaffold still alive, his heart was cut from his body and held aloft by the executioner with the cry: 'Here is the heart of a traitor!' Witnesses swore that they heard Sir Everard give answer: 'Thou liest!'

At the trial Everard had frankly admitted his own culpability, but had pleaded that his guilt should not be passed onto his family—least of all his infant sons, Kenelm and John. The prosecutors sneered at this request and the Attorney-General, Sir Edward Coke, showing no mercy, quoted a passage from Psalm 109: 'Let his wife be a widow, and his children vagabonds'. Small wonder that Kenelm ended up living in France for long periods!

At some point after his return to England in 1639, following one of his lengthy sojourns in France, Digby encountered Gascoigne in York. The meeting probably took place in late September 1640 around the time of the emergency Great Council of Peers, which had been convened there by the King. Quite possibly, the meeting was pre-arranged, since the young astronomer had brought along some of his inventions and the mutual friend, who introduced the two to each other, urged him to demonstrate them.

After showing Digby the essential Keplerian telescope arrangement, Gascoigne proceeded to explain how a scale could be inserted into the tube precisely where the real image was formed—thus permitting the image and scale to be both in focus and easily compared. As could be expected, Digby was much taken with the novel device and asked Gascoigne to forward a more detailed explanation, with diagrams and examples of some of the pioneering observations made with it.

Gascoigne duly obliged and his letter of 30 November 1640 was the result.

In the letter he is at pains to explain that the arrangement which had excited Digby's interest in York is simply a prototype and the enclosed observations are but a foretaste of what might be achieved. The possibilities, if higher quality components could be procured, are what interest him.

> Those observations of the Sun & Starres here denoted are not by me believed to be true <to> a second though I am confident of the possibility of that by a glasse no longer then that you did see at Yorke if the scale, case, & glasses were carefully made by skillfull artists. When I have the last according to my owne desire I can readily procure the other ...[12]

Alive to practical applications, he mentions how the new instrument could serve to measure the relative positions of celestial bodies. As a particular case, it might at last be possible, by accurately measuring the Moon's distance from specific stars, to deduce the terrestrial longitude of an observer. The basic method had been proposed by Johann Werner in 1514: The first step would be to measure the distance of the Moon from fixed background stars; Then, one would compare this with tables showing the time at which the Moon should have reached this position, if observed from a reference location (the place where the tables had been drawn up). Because the

[12] *Ibid*, f.54r.

Earth is spinning on its own axis through an angle of 15 ° every hour, the difference (in hours) between the *table* time and the *local* time when the observation was actually made, if multiplied by 15, will give how many degrees the observer is to the west of the reference location.

In truth, Werner's method was impracticable—not only because a precision instrument, such as Gascoigne's, was lacking, and adequate lunar tables, but also because Werner had not taken into account the effect of lunar parallax on measurements of the Moon's position.

The French astronomer Jean-Baptiste Morin (1583–1656) corrected the parallax flaw in the lunar distance method. Morin was eager to capitalise on his innovation. However, this was to land him in disputations with other astronomers as he single-mindedly sought to obtain a reward in the form of a pension from Cardinal Richelieu. In the letter to Digby, Gascoigne speaks of a further enhancement to his micrometer and speculates on how useful it might have been to the Frenchman:

> I have an other additament that will afforde the inclination of one star to an other in any verticall of the excellent use for the invention of their longitudes: How far Morinus would have stretched such an invention for his owne, not Countries profit had he knowne it his book witnesseth.[13]

Gascoigne mentions in the letter that he hopes lunar observations with his micrometer will help 'to evince the Moon's distance from the Earth onely from its spots'.[14] The plan is not elaborated any further here, but over the next few months it would feature in his correspondence with William Crabtree. 'I also hope to find', he wrote to Crabtree on 2 December, 'the Moon's parallax of altitude in the meridian by its owne body, though Morinus thinke it impossible'.[15] The proposal was to measure the angular separation of a spot (presumably a crater) from the limb of the Moon when the Moon was on the horizon and when it was on the meridian. The displacement of the observing sites (equivalent to the radius of the Earth) would lead to a difference in apparent separation. Hence, knowing the Earth's radius, the distance of the Moon—or its parallax—could be calculated. In a further letter to Crabtree on 25 Jan 1641 he returned to the same idea. We don't have Crabtree's response, but in a note accompanying his transcription of this letter, Flamsteed says: 'his method is demonstrable, but as Mr. Crabtree proves not practicable: by reason that an error of 2 or 3″ in observation is inevitable which will destroy the fruit of the operation and distort it beyond the granted truth'.[16]

The letter to Digby then mentions briefly some further applications that Gascoigne foresees for the micrometer:

> Each dayes alteration of [the Moon's] diameter is as conspicuous as the encrease or diminution of its light. Neyther are the other planets priviledged by their very acute angles, nor need Galileus proiect of a hill & Collumne or Cilinder looked at at divers miles distance;

[13] *Ibid*, f.54r.

[14] *Ibid*, f.54r.

[15] Bodleian Library, MS.Eng.7031, f.44r.

[16] The National Archives, Flamsteed papers, RGO 1/40, f.13v.

> but may by a large & good perspective glasse have their diameters defined in some <par>ts of seconds or decimalls answerable as I have experimented on Mars …[17]

Measurement of subtle changes in the apparent lunar diameter will be possible—thus enabling the shape of the orbit to be finally resolved. The planets too will yield their diameters to the mechanism of the micrometer. Gascoigne also understands that the key to proving the Copernican theory also resides in refinements of the same instrument. This is the meaning of his reference to Galileo's project.

Through the character of Salviati, in his *Dialogue Concerning the Two Chief World Systems* (1632 in Italian, 1635 in Latin), Galileo had postulated that a distant mountain and the beams forming part of a chapel at the summit could be used—in conjunction with a telescope—as a marker to detect the annual parallax of the stars and hence prove the heliocentric theory.[18] Gascoigne assures Digby that his micrometer, when fitted to a more perfect telescope than the prototype that he has already tested on Mars, will be far more convenient for the task.

In explaining the trigonometry behind the refraction of rays through a plano-convex lens, Gascoigne took care to go into some detail, but was wary of seeming to 'talk down' to such a learned scholar as Digby: 'My desire to be understood prevails with me though I know you litle neede it to adde one line of explication'. He would use exactly the same phrase with Crabtree a few days later,[19] but knew that his ideas were so novel that inevitably anyone who wanted to understand them would have some supplementary questions. Therefore, he advised diplomatically later in the letter: 'If in any part I be not understood upon your information of me I shall be ready to render it more at large'. The new national postal service, established by royal proclamation in 1635, would provide (at a cost of 6d per letter[20]) a convenient conduit to Middleton: 'If at any time you please to write, a letter left eyther at the Beare in Basing Hall Street or sent to the Post that comes to York will not faile of being delivered if directed thither whence this is dated'.[21]

Finally, Gascoigne entreats Digby to bring any relevant new books to his attention: 'If any mathematicall novelties come to your notice (which no rare book can escape) that I may be onely informed of their worth & subject shall no litle augment that obligation whereby I am bound…'. In the absence of journals, learned societies and and institutional libraries, such correspondence would be an essential means of keeping 'up to speed'.

---

[17] British Library MS, Add. 61873, f.54r.

[18] Galileo, *Dialogue concerning the two chief world systems*, transl. by Stillman Drake (Los Angeles, 1967), 387–9. In his second letter to William Oughtred in February 1641 (Rigaud, *op. cit.*, 48), Gascoigne refers to the same plan as "Galileo's prodigious project" and again says that his micrometer will render the plan unnecessary.

[19] The National Archives, RGO 1/40, f11r, letter from Gascoigne to Crabtree (2 December 1640).

[20] Beale, Philip, *Englands Mail: two millennia of letter writing* (Stroud, 2005), 207.

[21] The National Archives, RGO 1/40, f.54r.

Not surprisingly, no supplementary questions or book notifications from Digby have come to light. Maybe none were sent. Maybe the letter never even reached him. There would be a good reason for any of these eventualities.

Within 2 months of Gascoigne's letter, Kenelm Digby had been hauled to the bar of the Long Parliament's House of Commons to face an inquisitorial grilling by the *Select Committee for Popish Recusants and the Catholic Contribution.*

When Digby had arrived in England from France he had been approached by Queen Henrietta Maria with a request that he help raise money from the Catholic community to help pay for an expedition by Charles I against the Scots. He had already heard with alarm from one of Cardinal Richelieu's chaplains that anti-royalist, anti-Catholic forces numbering above 30,000 armed men were massing in Scotland and were poised for action. On 17 April 1639 the Queen signed a letter asking for a Catholic contribution to finance the King's army of the north to confront the threat and this was duly accompanied by supportive letters from Sir Walter Montagu and Digby.

Parliamentarians were outraged, deeming the King's forces as nothing less than a 'popish' army. Digby was summoned to appear before them and on 2 January 1641 the Select Committee was instructed 'to prepare questions for Digby and Montagu, etc., concerning motives and instructions for raising the recusant contribution for the northern expedition'.[22] Although Digby, as usual, appeared disarmingly open when questioned, the upshot was that he was eventually imprisoned by the Long Parliament and was not released until July 1643, after which he went into exile.

It cannot have escaped Gascoigne's notice that Digby was an embattled man—mired in political controversy. Indeed, much of the talk at the Great Council in York must have centred on the usefulness, or otherwise, of the 'northern expedition'. Digby's stance must have been immaterial to Gascoigne, or entirely in keeping with his own sentiments.

Talk of the developing political crisis was pervading every stratum of society and the consequent tensions were placing an intolerable strain on social bonds. Every subject of the King, however disinclined to become embroiled, was inexorably pushed to take sides.

[22] Jansson, Maija, *Proceedings in the Opening Session of the Long Parliament, House of Commons*, v.2: 21 December 1640–1620 March 1641 (Rochester, NY, 2000–2007), 294.

Part IV
The Civil War and After

Chapter 14
The Road to Civil War

In 1883, according to the *Palatine Note-Book* of that year, there took place 'one of the most deplorable dispersions of an old family library at Sotheby's … when the valuable Towneley manuscripts and the remains of the noble library chiefly collected by Richard Towneley, the philosopher, were sold by his descendants'.[1] From this sale, Chethams Library in Manchester—Britain's oldest surviving public library—acquired a seventeenth-century notebook that had been listed in the sale catalogue as '*A new and true theory of the sun's motion. MS*'. It is now catalogued as 'Notebook [by William Gascoigne (1612–1644)] containing notes in English taken from the 1632 edition of Lansberge's *Tabulae Motuum Coelestium*'.[2] The notebook is 118 pages long and largely consists of a verbatim English translation of Lansberge's '*Novae & genuinae Motuum Coelestium Theoricae*', which was appended to the 1632 edition of *Tabulae Motuum*. The rear of the first page of the notebook has several attempts at a latinised version of Gascoigne's signature (Fig. 14.1), but these are dissimilar to the uniform signatures on the three known Gascoigne letters. The handwriting throughout the notebook, although similar to that of the letters, does have some important differences. Being a verbatim translation, the notebook casts no fresh light on the scribe's astronomical ideas, but the extensive list of aphorisms written in a (possibly) more juvenile hand on the first few pages do show the dark side of the contemporary psyche (Fig. 14.2). A few extracts will give the flavour:

> Traitors are odious to those whose instruments they be.
> Hatred is the more deadly by how more unjustly deserved.
> The sight of the dead unburied will make soldiers more fearful and less forward to fight.
> We should rather die than suffer our Country to be brought to bondage.
> It is the part of a coward to fly, if his slaying or death might profit his country.

The earnest study of positional astronomy, typified by the greater part of the notebook (Fig. 14.3), dominated Gascoigne's known correspondence for a period

[1] *The Palatine note-book* (Manchester, August 1883), iii, 187.
[2] Chethams Library, *Manuscripts/1/227*, Finding No. A.3.110.

D. Sellers, *In Search of William Gascoigne: Seventeenth Century Astronomer*, Astrophysics and Space Science Library 390, DOI 10.1007/978-1-4614-4097-0_14,

Fig. 14.1 A juvenile signature of William Gascoigne in the Chetham's notebook? (Courtesy of Chetham's Library, Manchester)

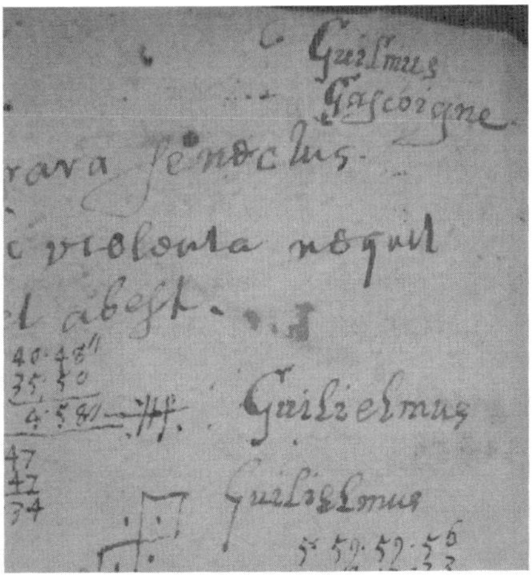

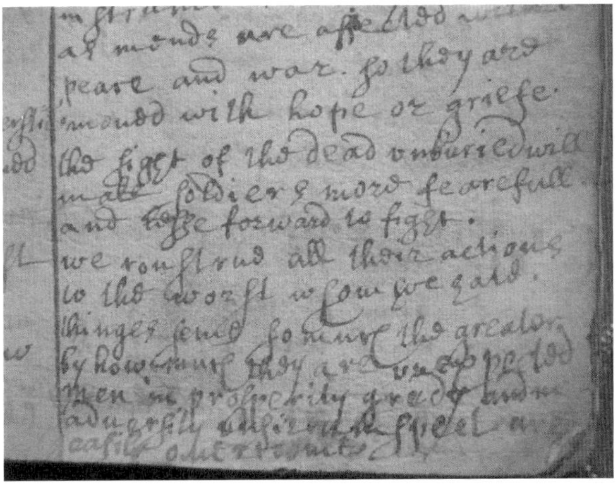

Fig. 14.2 Sombre aphorisms in the Chetham's notebook (Courtesy of Chetham's Library, Manchester)

of scarcely 3 years before he was swept up in the maelstrom of a civil war in which these aphorisms were the fearful daily currency.

As the spring of 1642 arrived, Gascoigne awaited the delivery of his freshly ordered 5-ft iron quadrant. Eagerly, he anticipated the prospect of measuring celestial angles to one-thousandth part of a degree. But events 200 miles away were conspiring to bring an end to his astronomical pursuits.

On 4th January 1642 King Charles I had provocatively entered the House of Commons, trying to arrest five of his most determined opponents.[3] The Commons

[3] Pym, Hampden, Haselrig, Holder and Strode.

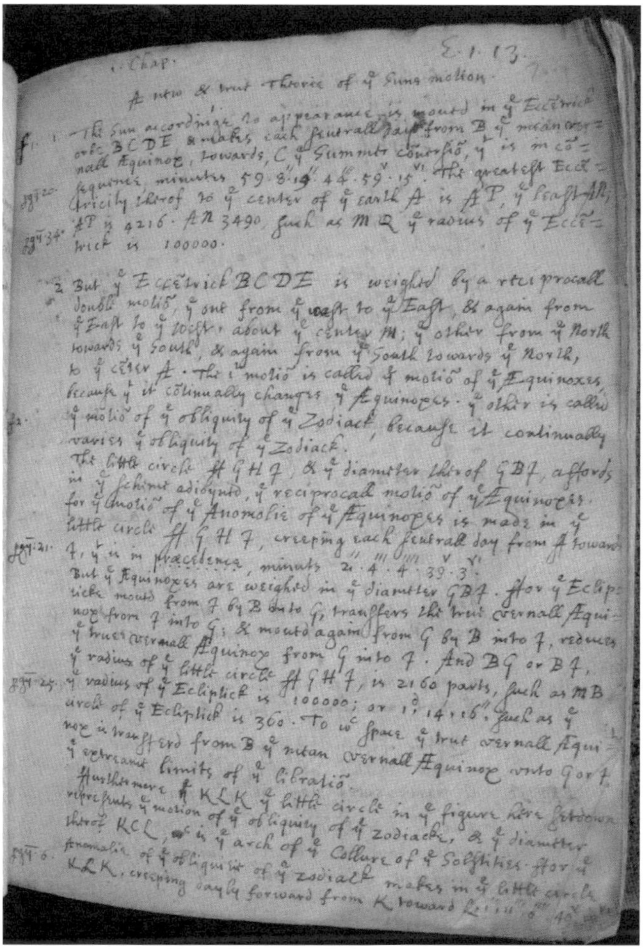

Fig. 14.3 A sample page from the Chetham's notebook (Courtesy of Chetham's Library, Manchester)

stood firm against this attempted coup. Within 3 months Charles was forced to flee from the capital. The next time he would enter London would be when on trial for his life.

After fleeing London, the King headed for York, the most important royalist garrison town in the North, and here set about gathering supporters and preparing for war. Likewise, the parliamentary side increased its combative stance and mustered its forces. In early March 1642 the Houses of Commons and Lords resolved that the kingdom should be put 'in a posture of defence' and passed the Militia Ordinance, giving itself, rather than the King, authority to command the armed forces. The King responded by declaring obeisance to the Ordinance as treasonous.

Consolidating their demands, both Houses of Parliament on 1st June 1642 approved 19 Propositions, which were then sent to the King at York. Moderates among the parliamentary ranks—mistaking the significance of their own actions—regarded the propositions as a basis for negotiation. Others regarded the propositions as an ultimatum.

The key demands contained in the text submitted to the King were that the King's ministers must be answerable to Parliament, that Parliament must be responsible for the defence of the country (the King would have to seek its authority to raise arms), that Parliament would supervise foreign policy, and that stringent new laws against Catholics ('Jesuits, priests and Popish recusants') must be enforced.

The King flatly refused to accept the Propositions, characterising them as 'a total Subversion of the Fundamental Laws, and that excellent Constitution of this Kingdom, which hath made this Nation so many years both Famous and Happy'. The encroaching of Democracy upon the rights of Absolute Monarchy and Aristocracy would ultimately, he claimed, 'Destroy all Rights and Proprieties, all distinctions of Families and Merit; And by this means this splendid and excellently distinguished form of Government end in a dark equall Chaos of Confusion, and the long Line of Our many noble Ancestors in an Jack Cade, or a Wat Tyler.'[4]

In short, his answer was 'We are unwilling to change the laws of England'.

The manoeuvrings of both sides eventually gave way to outbreaks of fighting between parliamentary and royalist supporters and minor military skirmishes, in which dozens died.

On the 22nd August 1642 the King raised the Royal Standard at Nottingham Castle and declared war on Parliament. The Civil War had begun.

Looked at from the safe distance of several centuries, the English Civil War can be seen in terms of simple analytical themes: as a struggle of republicanism against monarchy, of burgeoning bourgeois democracy against the feudal order, of puritanism and 'the true Protestant religion' against 'Popery', of freedom against tyranny. To the participants, however, and to the population at large, the issues at stake were far from simple.

For instance, although the 19 Propositions insisted that the laws in force against Catholics be executed 'without any toleration or dispensation to the contrary', this did not automatically push all Catholics headlong into the royalist camp. Indeed, Charles made many anti-Catholic utterances and in his Declaration from Nottingham of 25th August 1642, he assured potential supporters that 'Nothing shall be wanting on our part which may advance the true Protestant religion, oppose Popery and superstition'.[5]

In truth, the approaching conflict—eventually enveloping every corner of the land—would turn neighbour against neighbour, and friend against friend. According to the Gloucestershire nonconformist, John Corbet, 'fears and jealousies had so

[4] Jack Cade was an English rebel who led an unsuccessful rebellion against Henry VI in 1450. Wat Tyler led a peasants' rebellion against Richard II in 1381.

[5] Daniels, C.W., and Morill, J., *Charles I* (Cambridge, 1988), 101.

possessed the kingdom that a man could hardly travel through any market town but he should be asked whether he were for King or Parliament.'[6]

It was inbuilt in the royalists' *commision of array* and the parliamentary *Militia Ordinance*, according to civil war historian P.R. Newman, that the first call to duty would go, via the lieutenants of counties, to the leading landowners and thence it would percolate to such as the Gascoignes and the various strata of the gentry:

> They might be largely inexperienced in military matters; they might be wary of committing their resources and those of their neighbours to a thing so wasteful as civil war, but the whole structure of society, the whole ethos of their order, required from them such sacrifices as part of the price of their status, and in the light of a risk of social disintegration, they would respond despite their personal reservations.[7]

We will perhaps never know what motivation impelled William Gascoigne to join the King's side in the conflict: Whether it was in defence of his religion, defence of his family's economic interests, or some other imperative is unclear. How much did pressure from relatives and peers figure in his decision? His uncle and neighbour, Sir Ferdinando Leigh was a fervent monarchist and commander of a royalist regiment. Whatever the case may have been, at some stage following his last recorded letter to Crabtree, in June 1642, Gascoigne abandoned his telescopes and threw in his lot with the Cavalier army of Charles I.

The record of his role in the Civil War is uncertain. Historians give conflicting accounts of his activities. Many of these accounts are unreliable because they are based on contemporary reports of individuals with a similar surname who cannot be demonstrated conclusively to be William Gascoigne of Middleton.[8] The statement by Sir Charles Cavendish that Gascoigne was a 'providore' to the royalist army seems to be the only account that is unambiguously referring to the astronomer.

By the first half of 1644 the civil war had reached, possibly, its most decisive stage. Both sides were locked in a seemingly intractable conflict: Both hoping for outside assistance in order to tip the balance. The Royalists were banking on the King's army coming from Ireland. The Parliamentarians were counting on the arrival of Scottish troops having a decisive effect. The North of England was the battleground upon which these two armies of intervention, along with their allies,

---

[6] Adair, John, *By the Sword Divided* (Stroud, 1998), 22.

[7] Newman, P.R., *The Royalist Officer Corps 1642–1660: army command as a reflexion of the social structure*, The Historical Journal, xxvi (1983), 947.

[8] Charles Webster states that "William Gascoigne was a captain in Sir Marmaduke Langdale's cavalry regiment", *Richard Towneley (1629–1707), the Towneley group and seventeenth-century science, Transactions of the Historic Society of Lancashire and Cheshire*, cxviii (1967), 61; PR Newman says that a William Gascoigne was a "volunteer/captain" in Sir John Ramsden's Division, *The Royalist Army in Northern England*, D.Phil Thesis (University of York, 1978), i, 448; W.Wheater says "In his account of the first seige of Pontefract Castle in December 1644, Drake enumerates among the gentlemen volunteers under the command of Sir John Ramsden, a Mr William Gaskon, but whether he is the same William Gascoigne I cannot say", *Gentleman's Magazine* (London, 1863), 761; Taylor states that "in all probability [he] was one of the volunteer defenders of Pontefract Castle during the first seige", *Biographia Leodiensis*, 86.

fought for supremacy. It was clear that whoever won the battle for the North would control the outcome of the war.

Mid-summer of 1644 saw the fortunes of both sides finely balanced. York, the key to the North, the foremost city in a county with important ports and links to the continent, was occupied by Royalists under the command of William Cavendish, the Marquis of Newcastle. The Parliamentary army of Fairfax was laying siege at the gates.

On the eve of his entry into York with about 7,000 men, on 19th April, Newcastle had warned King Charles that the Royalist position in Yorkshire was in great peril: '… the Scots and Fairfax [Ferdinando, Lord Fairfax, commander of the Northern Parliamentary forces], having joined near Wetherby, are now too strong for us in matters of the field … they have already put themselves in such a posture as will soon ruin us, being at York, unless there is some speedy course taken to give us relief, and that with a considerable force, for their army is very strong … We shall be distressed here very shortly.'[9]

The attempt by Prince Rupert, the Kings nephew, to provide the necessary relief was to lead to one of the bloodiest battles ever fought on English soil. Rupert, accompanied by a force of approximately 7,000 cavalry, 6,000 foot soldiers and a reputation for invincibility, though only in his mid-twenties, marched towards York from the west. They reached Knaresborough (about 27 km from York) on 30th June. Ahead of them, the Parliamentarians, informed of Rupert's advance, temporarily abandoned the seige of York and during the night of 30th June/1st July set up position across the York to Knaresborough road, ready to intercept the Royalists. Under the command of the Earl of Leven more than 21,000 men of the Parliamentary army and their Scottish allies lay in wait north-west of the tiny village of Long Marston, 11 km to the west of York. But Rupert had no intention of facing such a large force in battle until he had linked up with Newcastle's troops—currently in York.

On 1st July, misleading the enemy with respect to his true intentions, Rupert dispatched a sizeable force of cavalry up the main highway towards Long Marston. The Parliamentarians, believing that this was the prelude to a bitter struggle, made ready for battle. It was not until some considerable time had elapsed, with no sign of a further move from the Royalists, that Leven realised he had been deceived. The bulk of Rupert's army had made a large detour to the north-east and had effectively passed behind the Parliamentary ranks to gain the unguarded Boroughbridge-to-York Road and thence, with minor difficulty, had been able to relieve York.

The time of reckoning was now at hand. Rupert wasted no time. Instead of entering York, he issued instructions that the Marquis of Newcastle should bring his troops out of York and meet up with those of Rupert on Marston Moor by 4 a.m. the next morning.

As dawn broke on 2nd July 1644, cavalry units of the Royalist army were already making their way across the River Ouse, by means of a bridge of boats, and heading for Marston Moor. Rupert was determined to seize the advantage and drive the

[9] Barratt, John, *The Battle for York – Marston Moor 1644* (Stroud, 2002), 42.

Fig. 14.4 Marston Moor. The land in the foreground was occupied by the royalists, whilst the ridge in the distance was the location of the parliamentarian encampment (Author's collection)

Parliamentarians off the ridge overlooking the Moor (Fig. 14.4), whilst their battle plans were still in disarray. The alacrity of Rupert, however, was seemingly not matched by any sense of haste back in York. By the time sporadic skirmishing commenced, at around 9 a.m., Newcastle and his forces were only just leaving the city.

Without Newcastle's reinforcements, Prince Rupert felt unable to take the offensive in a determined fashion. Impatiently he was all the time 'sending away to my Lord Newcastle to march with all speed.'[10]

Throughout the remainder of the morning the two opposing armies gathered their forces: the Parliamentarians along the ridge to the south of Marston Moor; the Royalists on the flat plain of the Moor itself. The opportunity presented by the initial decisiveness of Rupert was rapidly vanishing. By mid-afternoon the army deployed on the ridge totalled about 20,000 soldiers. By contrast, the Royalist ranks below comprised 17,000 – including 4,500 horse. As evening approached, Rupert resolved to leave an attack until the following day. He felt sure that the enemy too would defer the reckoning. Indeed, for the time-being, apart from random musket shots, all that could be discerned in the opposing camp was the singing of psalms.[11]

[10] *Ibid.*, 76.

[11] Slingsby, Sir Henry, *Diary of Sir Henry Slingsby*, quoted by Peter Young, *Marston Moor: The Campaign & the Battle* (Kineton, 1970), 216.

Fig. 14.5 A re-enactment of the Battle of Marston Moor by the Sealed Knot (Author's collection)

Now it was Rupert's turn to be deceived. As he and his forces relaxed and started about their evening meal, the Parliamentarians decided to launch a general attack. One of their number, Lionel Watson, Scoutmaster-General to the Earl of Manchester, later recalled 'about halfe an houre after seven a clock at night, we seeing the enemy would not charge us, we resolved by the help of God, to charge them, and so the signe being given, we marched down to the charge …'[12]

Soon intense fighting was to be seen along a 3 km battlefront. By which time, in the words of one Parliamentary captain recalling the fray, 'the main bodyes joining, made such a noise with shot and clamour of shouting that we lost our eares, and the smoke of powder was so thick that we saw no light but what proceeded from guns.'[13] (Fig. 14.5)

At first it appeared that the Royalists had the upper hand. Within little more than an hour they were gaining much of the ground previously occupied by their enemy, many of whose troops had taken flight. Through the agency of soldiers thus escaping from the scene of carnage reports and rumours spread of a great Royalist victory. The Royalist newspaper, *Mercurius Aulicus*, trumpeted that 'the great news of all is of what's done at Yorke … that the rebels are absolutely routed, that the Prince with his Army hath taken 48 peece of cannon, General Lesley and Sir Thomas Fairfax prisoners.'[14] Such accounts, however, proved to be decidedly premature.

Forces under the command of Oliver Cromwell and the Earl of Manchester launched a determined counter-attack and before midnight the situation had been reversed. The Royalists were vanquished and on the run. More than 4,000 men

[12] Barratt,John, *Op.Cit.*, 100.

[13] Young, Peter, *Op.Cit.*, 247.

[14] *The English Revolution III, Newsbooks 1: Oxford Royalist*, v.3 (London, 1971), 1072 (Mercurius Aulicus, 6 July 1644).

perished on Marston Moor (Fig. 14.6): most of them from the Royalist ranks. Simeon Ashe, one of Manchester's non-combatant chaplains, described the appalling aftermath: 'That night we kept the field, when the bodies of the dead were stripped. In the morning there was a mortifying object to behold, when the naked bodies of thousands lay upon the ground, and many not altogether dead.'[15]

The roads between the battle site and York were crowded with the fleeing remants of the defeated Royalist army, but the city gates were closed to prevent their pursuers from entering the city. Hence, as Sir Henry Slingsby witnessed, 'At the bar [Micklegate] none was suffered to come in but such as were of the town, so that the whole street was thronged up to the bar with wounded and lame people, which made a pitiful cry among them.'[16]

Amongst the dead bodies strewn across Marston Moor was that of Colonel Charles Towneley. His brother, Christopher was amongst those captured and imprisoned. Both had been with the Lancashire army that accompanied Prince Rupert on his journey over the Pennines, bound for York. Thus the fraternal partnership which had been pivotal in cementing communications between the exponents of the nascent new astronomy in England came to an end.

Towneley Hall at Burnley had been taken over by Parliamentarians at the start of the war and during the recent conflict Mary Towneley, the wife of Charles, had been staying in Knaresborough at the home of her father—Sir Francis Trappes. Upon hearing that her husband had died in the battle, she journeyed the next day to Marston Moor to recover his body. As she wandered disconsolately about the field with its distressing scenes, she was approached by a senior officer from the victorious side. He listened with sympathy to her plight, but urged her to leave the field lest she be molested by remaining soldiers. She reluctantly agreed and the officer summoned a trusted Roundhead trooper to carry her on his horse to a place of safety. As they rode away, she asked the trooper for the name of the officer who had shown such unexpected sympathy and kindness. 'It was Oliver Cromwell', came the reply.

Many later accounts report that William Gascoigne also perished in this bloody battle. In the *Astronomical Supplement* to his 1675 translation of *The Sphere* by Marcus Manilius, Sir Edward Sherburne wrote that Gascoigne 'and all his excellent Inventions and labours were lost by his Death, which was in his late Majesty's Service, in the Fight at Marston Moore.'[17] One of the tragedies of the English Civil War, however, was that there was often no way of knowing when, where or how a friend or relative had perished. Those who survived the ordeal of battle in many cases could not return home with their stories, but had to exile themselves. Sir Henry Slingsby, who lived but a few kilometres from the battlefield spoke of this predicament as he slipped away from the area: 'Thus disconsolate we march, forced to leave our country, unless we would apostate, not daring to see mine own house, nor

---

[15] Cooke, David, *The Civil War in Yorkshire: Fairfax versus Newcastle* (Barnsley, 2004), 148.

[16] Cooke, David, *Op.Cit*., 145; Barratt, John, *Op.Cit*., 141; Young, Peter, *Op.Cit*., 216.

[17] Sherburne, Sir Edward, *The Sphere of Marcus Manilius made an English Poem with Annotations and an Astronomical Appendix* (London, 1675), 92 (Appendix).

take a farewell of my children, although we lay the first night at Hessay within 2 miles of my house.'[18]

No truly contemporary reports mention the death of Gascoigne (or Captain Gascoigne, as some would have it) on 2nd July 1644 at Marston Moor, near York, but this should not surprise us. As P.R. Newman starkly notes,

> Colonels and lieutenant colonels tended to be noted as they fell in action, or as they died of wounds or privation … Majors were less often noted. Captains passed from the scene almost without comment, whilst lieutenants, ensigns, cornets and quartermasters were heaped anonymously with troopers and infantrymen. Full burial pits on many a battlefield contained, still contain, tumbled together, the naked corpses of officers, gentlemen and rank and file.[19]

Writing some 7 years after Sherburne, John Flamsteed—the first Astronomer Royal—concluded that Gascoigne 'was slaien at Yorke fight'[20] In later correspondence Flamsteed says that 'having engaged on the King's side he was slaine in his service at Marston Moore fight July the 2. 1644 sometime after Mr. Christopher Towneley of the Car in Lancashire.'[21] The reference to *Christopher* Towneley is plainly a slip of the pen, since Christopher survived the battle and lived until 1674. What is more, Christopher himself was apparently the direct source of Flamsteed's information. A rarely cited note in Flamsteed's own papers says 'Mr. Christopher Towneley informes mee that Mr. William Gascoigne was slaine at Yorke fight by a bullet as I remember. & that himselfe stood near if not next him when he fell'.[22]

Flamsteed's source seems to be impeccable—as near to an eyewitness as one could get. Nevertheless, elements of doubt persist as other researchers—including Gascoigne's contemporary and co-parishioner John Hopkinson—suggest that Gascoigne did not die at Marston Moor, but was killed a full 4 months earlier in a conflict at Melton Mowbray.[23] At the beginning of March 1644, the royalist newsletter *Mercurius Aulicus* reported that the forces of Sir Marmaduke Langdale had routed parliamentary troops outside Melton on 25th February.

> The Enemy was advanced through the Towne to meet us, and in a gallant fury gave us a bold charge upon the very ground chosen by our selves to fight upon; the encounter continued hot and sharp a good while, with severall various appearances of successe on both sides; but

[18] Cooke, David, *Op.Cit.*, 151.

[19] Newman, Peter Robert, *Op.cit.* (ref. 89), i, 38.

[20] Forbes, E.G., Murdin, L., and Willmoth, F., *The correspondence of John Flamsteed, the first Astronomer Royal* (London, 1995), i, 897, Letter 450, Flamsteed to Molyneux, 29 May 1682.

[21] Forbes, E.G., Murdin, L., and Willmoth, F., *The correspondence of John Flamsteed, the first Astronomer Royal* (London, 1997), ii, 421, Letter 622, Flamsteed to Molyneux, 10 May 1690.

[22] The National Archives, *Flamsteed Papers*, RGO, 1/40, f.22r. Flamsteed must have obtained this story from Christopher Towneley between 1672 and 1674, since on 13 August 1672 he wrote to John Collins that Gascoigne must have died in 1642. Towneley, the source of the revised version, died in 1674.

[23] Hopkinson, John, *A Collection of Ye Pedigrees and Descents of severall of the Gentry of the West Riding of the County of York and Elsewhere*, 446 (Gascoigne of Thorp on ye Hill). A copy is held at Leeds Central Library (Phillips MS. 11948).

Fig. 14.6 The memorial to the fallen at Marston Moor (erected by the Cromwell Association) (Author's collection)

> at length they were wholly routed, many of their Commanders slaine, many hurt, and all the body scattered unto their severall Garrisons, the fatigue of our long March, and the night intervening, prevented our very farre pursuit of them in a strange Country but we flew upon the place neere 100, and tooke almost so many Prisoners and foure Colours of Horse. The losse we sustained was small, only Sir John Girlington and Captain Gascoigne, two gallant Gentlemen, both slaine in the first charge, and some few wounded.[24]

It is not clear whether this is the same Gascoigne, so we may never be certain whether this 'small' loss to Langdale's army was indeed the loss of the astronomer, whose demise was such a blow to English science.

[24] *The English Revolution III, Newsbooks 1: Oxford Royalist*, v.3 (London, 1971), 1402–3(Mercurius Aulicus, 8 March 1644).

Chapter 15
After Marston Moor

After the catastrophe at Marston Moor, the fortunes of the Royalists rapidly waned. Decisive defeat at Naseby in June 1645 was followed by last ditch stands at heavily besieged outposts, not least on the Gascoignes' home terrain in Yorkshire.

In his *Journal of the first and second seiges of Pontefract Castle (1644–1645)*, Nathan Drake, a 'gentleman volunteer therein', gives a list of all the volunteers that were at the castle on 25th December 1644. This includes a 'Mr. Gaskon', under the heading 'Gentlemen Vollunteres'. The same individual is later referred to as 'Mr. Gascone', a companion of Drake in Sir John Ramsden's division.[1]

Some references to Gascoignes at these and later sieges might be related to the activities of William's father, Henry, who was almost certainly also drawn into the conflict. On 16 November 1643, Henry Gascoigne was sent 'Directions from his Majesty's Commissioners of York concerneing the raising of moneyes & Provisions for the army under command of the marquesse of Newcastle'.[2] Since the directions related to 'Agbridge & Morley' and Middleton falls into the Agbrigg wapentake (an administrative district), it seems reasonable to assume that this is addressed to our Henry—i.e. Henry Gascoigne of Thorp-on-the-Hill. The other addressees are: Thomas Stringer, John Hodgson, Henry Favill, John Hobkinson, Abraham Sunderland and Thomas Thornehill. The first two were amongst the gentlemen volunteers at Pontefract Castle and Abraham Sunderland died there.

The directions advised the seven men that a warrant had been issued to Chief Constables requiring them to impose a monthly levy on the inhabitants of local

[1] Drake, Nathan, A *Journal of the first and second seiges of Pontefract Castle (1644–1645)*, Publications of the Surtees Society, xxxvii (Durham, 1861), Miscellanea.

[2] Northamptonshire Record Office, *Westmorland (Apethorpe) Collection*, ref. W(A) box 7/no.13. A bundle marked "Old letters and copys of Old Title Deeds" contains an item catalogued as "Directions for Thomas Stringer, Henrye Gascoigne and Jno. Hobkinson and others concerning the raiseing of moneyes and Provisions for the armye under comand of the Marquesse of Neucastle dated November 16, 1643 from his Majesties Commissioners at Yorke".

D. Sellers, *In Search of William Gascoigne: Seventeenth Century Astronomer*, Astrophysics and Space Science Library 390, DOI 10.1007/978-1-4614-4097-0_15,

townships, part of which was to be the provision of victuals and part in the form of money, to be paid to Sir William Carnabie, 'Treasurer at warrs'.

> Now this much is desired from you, that you will call upon & hasten the Chief Constables .... to charge the severall Townes .... You are also desired to divide the Townes within the division among your selves, takeing each one of you a certaine Number of Townes to your particular care & charge .... Then to see that money & provisions due from those Townes .... be duely & orderly payd to you.
>
> It is desired that if any person shall faile to pay their moneys or provisions, contrary to your expectation, then you will use your best meanes to obtaine it, & for your aide theirin, you are desired to imploy Captaine Whittington & Capt. Benson their troopes of horse which doe attend you

The signatories to these directions were Edward Stanhope, Matthew Hutton, William Savile, Robert Rockley and Paul Neile. The latter was later to be one of the founding members of the Royal Society and an expert on the grinding of telescope lenses.

Whatever the success of Henry Gascoigne and his colleagues in collecting revenues for the King's cause, it ultimately counted for nought. In the year following the rout at Marston Moor one royalist stronghold after another fell to the parliamentary forces. Pontefract Castle endured many months of intense bombardment. Cromwell himself was to observe that it was 'one of the strongest Garrisons in the Kingdom; well-watered; situated on rock in every part of it; and therefore difficult to mine. The walls are very thick and high, with strong towers; and if battered, very difficult of access'.[3] Only starvation of the defenders caused its eventual surrender on 20th July 1645.

By this time, however, Henry had moved to Sandal Castle, near Wakefield and was therefore amongst the recipients of a jubilant ultimatum, swiftly dispatched by the victors at Pontefract:

> For the Commander in Chiefe of Sandall Castle, and the Gentlemen and Officers there.
>
> Gentlemen,
>
> You cannot be ignorant that Pontefract Castle is yielded unto us, and the Castle of Sandall having been verie vexatious to these parts of the Countrey, wee are constrained for the preservation of the well-affected people, who complaine much of their sufferings by it, to sit downe with our Forces before it, intending by God's assistance, not to rise thence till wee have carried that place: yet neverthelesse, because our resolutions are to yeeld honourable courses, to avoyd the effusion of Christian bloud, which if that place be taken by force, must of necessitie be shed on both sides. We have therefore thought it most convenient to send you this Summons...[4]

The royalists were offered the opportunity to surrender on 'honourable termes'. Within the day, a defiant response was drafted, bearing the signature of Henry Gascoigne and eight others:

> You must beleeve, that the taking of Pontefract Castle cannot take away our Allegeance, but shall contract it, and adde vigour to it ... For our feares of your power against us, they are

[3] Roberts, Ian, *Pontefract Castle* (Wakefield, 1990), 43.

[4] Walker, J.W., *Sandal Castle*, Yorkshire Archaeological Society Journal, xiii (London, 1895), 184.

Fig. 15.1 Sandal Castle (From Yorkshire Archaeological Journal, v.xiii (London, 1895))

> lesse than when you were remoter from us. This being read, you must credit our resolutions most peremptorie and unalterable, for the maintaining of this Castle against all those who shall summon it from any other authoritte than his Majesties Signet.[5]

It was a hopeless gesture for a lost cause. Sandal Castle (Fig. 15.1) was surrendered on 1st October 1645. Eighty or so soldiers were found inside, along with 150 muskets, 20 halberds, 140 swords, 2 barrels of gunpowder and a 'good store of provisions of victual'. Henry, must have left before this bitter end, since he was buried at Rothwell on 26th September 1645. Did he die of injuries sustained in the Sandal defence or of some other affliction? It is not possible to be certain until more evidence surfaces. His death could well have been caused by the plague, which ravaged Leeds and its environs in 1645. In the words of one account,

> To the horrors of war at this calamitous period were added the calamities of pestilence. The plague broke out with tremendous violence in Leeds in March, 1645, and, by the twenty-fifth of December in that year, thirteen hundred and twenty-five persons, or more than one-fifth of the whole population, had perished. The disease was not confined to the closest and most densely inhabited parts of the town, but extended to its most open districts and its most airy suburbs. At the same time, its ravages were the most fatal in the streets and lanes, where the poorest, the worst fed, the worst clothed, and the worst housed people resided. So great was the consternation, that all who were able, fled from the scene of contamination and death; the grass grew in the deserted streets; the markets were removed to Woodhouse; the doors of the old church were closed; and, if the testimony of contemporaneous witnesses is to be credited, the very birds fell dead as they flew over the town.[6]

---

[5] *Ibid.*, 184.

[6] Parsons, Edward, *The Tourist's Companion; or, the History of Scenes and Places on the Route of the Railroad* (London, 1835), 10–11.

For the royalists who escaped the ravages of war and plague the outlook was bleak as the victorious party, classing the losers as 'delinquents', stepped up the sequestration of their property or exacted punitive fines for its return.

One of the preachers to the besieged royalist garrison at Pontefract Castle was Henry Gascoigne's co-parishioner: Edmund Kay, the Vicar of Rothwell. Kay paid dearly for his loyalty to the King. According to Walker's propagandist *Suffering of the Clergy arising out of the Civil War*, 'he was plundered while he had a stool to sit on; he and his family were turned out of doors, was carried prisoner from place to place and would have perished had it not been for the kindness and charity of Sir John Worsnam, of Nostell, who maintained him as long as he lived'.[7]

William Gascoigne's uncle, Sir Ferdinando Leigh of Middleton Hall, who had been a fanatical royalist participant in the conflict, survived but feigned madness in an unsuccessful attempt to avoid being fined by the new Government. He claimed that he could not even remember the event of the Civil War. The official record describes his testimony thus:

> 9 July 1650: He certifies that about 13 Feb 1643 he was in York ill, and with such pains in his head that he knew not what he did. That he does not remember the Yorkshire Engagement, nor any bonds, nor anything been demanded of him. Asks what sums he is charged with, and time to make answer, on account of his age and infirmities.
>
> 31 July: He not appearing, his estates to be seized and sold till the portion his paid.[8]

The ingenious astronomer's untimely death in the English Civil War, at barely 32 years of age, would have resulted in his remaining papers, books and instruments coming into the possession of his father, Henry Gascoigne. Henry, however, died intestate. There was a dispute over his estate, administration of which was ultimately handed to a Peter Richardson of the city of York.[9] Although an inventory of

---

[7] Batty, John, *History of Rothwell* (Rothwell, 1877), 57–58.

[8] Newbould, John F., *A History of the Township of Middleton in the Parish of Rothwell*, Part 1 (1066–1750), Leeds, 2008, typescript in Leeds Central Library (ref Q 942.81/NEW), 33.

[9] Administration of Henry Gascoyne, *Pontefract Act Book*, fol.186 (Borthwick Institute, York).

> On the seventh day of the month of July in the year of the Lord 1646 the administration of the goods, rights and credits of Henry Gascoyne Esquire late of Middleton in the parish of Rothwell in the diocese of York, deceased intestate as it is asserted, was granted by and through the decree of the court to Peter Richardson of the city of York, yeoman, the servant of Henry Brearey of the same city of York, gentleman, creditor of the said deceased, previously having been sworn. A bond was entered but no inventory was exhibited. And on the thirteenth day of January in the year of the Lord 1647 an inventory was exhibited of above £40 through a true, full and present oath. [trans. Dr Philippa Hoskin]

One might have thought that there would be a record of the Richardson-Brearey dispute in the Court Books: That there might even be an inventory contained in that record however brief. Nevertheless, the new digital catalogue of the 20,000+ records of the Court Book index for York (1300–1858) recently placed online by the Borthwick Institute contains no reference to this case. In any event, the naming of specific books and instruments in inventories seems not to have been a regular practice at York from around the middle of Elizabeth's reign. The most that one could expect in this regard would be a reference to the value 'in books' or 'sundry instruments'.

the estate was eventually exhibited to the court in January 1647, no record of it survives, because such records were only kept from 1680 onwards. We may never know the fate of William Gascoigne's belongings. Of the few letters saved by contemporaries, the majority have long since disappeared into oblivion.

According to Flamsteed, speaking of Horrox, Crabtree and Gascoigne,

> After all of their deaths Mr Christopher Towneley of the Carr in Lancashire & Sr Jonas Moore went to Mr Crabtree's house where his widdow gave them all his papers. These they secured and afterwards sorted & made up that collection of Mr Horrox & Crabtree's Letters & observations [....] published by Dr Wallis. Mr Gascoigne's papers fell to Mr Towneleys [....] who exchanged them for others with Mr Richard Towneley of Towneley in Lancashire.[10]

We must be thankful that the 25 year old John Flamsteed so assiduously copied out some of the content in the Spring of 1672 and that the Reverend William Derham, stung by a French ascendancy in optics, published some excerpts in 1717. Without these fragments and the three letters in William Gascoigne's own hand, we would probably be unaware of the tragic story of that remarkable young man.

[10] The National Archives, RGO 1/65, f.112v.

Chapter 16
The Legacy of William Gascoigne

The English Civil War, which divided even families by the sword, expunged not only the life of William Gascoigne and many of his compatriots: it also wiped out much of the *record* of their lives and circumstances. Some of the new material that has been examined above—sparse though it is—helps us to start rebuilding this record.

The surviving primary sources that have been exposed hitherto have dealt with, or have been assessed in relation to, a relatively narrow compass of Gascoigne's activities. The transcriptions by Derham were all aimed at demonstrating the priority of Gascoigne's invention and use of the telescopic sight and micrometer. The surviving letters in the young astronomer's own hand (to Oughtred and Digby within a space of less than 3 months) are aimed at seeking recognition of these instruments. The prime motive of John Flamsteed in making transcriptions of the Gascoigne-Crabtree letters was to extract the pioneering ideas on analytical optics and to cull as many useful observations as he could. Even though Flamsteed's copies did include material on broader issues, this has tended to be passed over by subsequent authors.

The wider correspondence shows that Gascoigne, despite his relative physical isolation from other astronomers, was nevertheless pursuing research that was addressing the key astronomical issues of the day. He was obtaining and reading critically the latest astronomical writings published on the continent. He was working hard to develop an analytical approach to telescopic optics and was using his ingenious micrometer to test current theories of the terrestrial and lunar orbits.

It is possible that—even before Towneley's revelation of the micrometer, in response to Auzout's claims—some English astronomers, such as Christopher Wren, got wind of the techniques pioneered by Gascoigne. But if Auzout and others had to re-invent the micrometer, does this mean that, ultimately, Gascoigne's work had no lasting impact? An answer was given by Flamsteed when he wrote to Molyneux on 21 January 1683, declaring,

> … as for Kepler and De Chartes [i.e. Descartes] I know not how to excuse them except you allow a small faileing or two on the account of theire great merits otherwayes, I must

D. Sellers, *In Search of William Gascoigne: Seventeenth Century Astronomer*, Astrophysics and Space Science Library 390, DOI 10.1007/978-1-4614-4097-0_16,

> confess to you I have not the patience to peruse any of theire writeings on this subject haveing benefitted my selfe more by a letter and figure or two of Mr Gascoignes with an houres serious thoughts then I could have done by ten spent on the persuall of any of theire writeings...[1]

Perhaps Gascoigne's most effective legacy was his contribution to the optical instruction of those who, a generation or so later, would dedicate their lives to the meticulous measurement of the heavens. Upon their efforts, much of the seventeenth century progress of science depended.

The hagiographic essay on Gascoigne in R.V. Taylor's *Biographia Leodiensis* (1865) lamented that 'little is known of William except the immortal inventions which resulted from the continued labours of his great mind' and that his micrometer 'would have at once rendered his name illustrious, had not his untimely end, and the melancholy circumstances which produced it, given others an opportunity of claiming the honour and receiving the measure of applause the invention so nobly deserved'.[2] If this book has contributed in some measure to a more rounded view of Gascoigne's activities as an astronomer, then its purpose will have been fulfilled.

---

[1] Forbes, E.G., Murdin, L., and Willmoth, F., *The correspondence of John Flamsteed, the first Astronomer Royal* (London, 1997), ii, 83, Letter 485, Flamsteed to Molyneux, 21 January 1683.

[2] Taylor, Richard Vickerman, *Biographia leodiensis: or, Biographical sketches of the worthies of Leeds and neighbourhood, from the Norman conquest to the present time* (London, 1865), 86–7.

Part V
Appendices

Chapter 17
Appendix 1 Description of Mr. Gascoigne's Micrometer

By Robert Hooke

Philosophical transactions of the Royal Society, v.ii, (1667), 195–7

In [Fig. 3.3], the fig. 1, 2, and 3, represent the several parts of this instrument; fig. 4, part of the telescope with the instrument applied to it, and 5 the rest on which the whole is supported.

Fig. 1 represents the brass box with the whole instrument, except only the moveable cover, and the screws by which it is fixed to the telescope. In this figure ***aaaa***, is a small oblong brass box, serving both to contain the screws, and their sockets or female screws, and also to cause all the several moveable parts of the instrument to move very true, smooth, and in a simple direct motion. To one end is screwed on a round plate of brass, ***bbbb***, about three inches over; the extreme limb of its outside being divided into 100 equal parts, and numbered by 10, 20, 30, &c. Through the middle of this plate, and the middle of the box ***aaa***, is placed a very curiously[1] wrought screw, about the size of a goose quill, and of the length of the box, the head of which is, by a fixed ring or shoulder on the inside, and a small springing plate ***dd*** on the outside, so adapted to the plate that it is not in the least subject to shake. The other end of this screw is, by another little screw (whose small point fills the centre or hole made in the end of the longer screw for this purpose) rendered so fixed and steady in the box, that there appears not the least danger of shaking. On the head of this screw, without[2] the springing plate, is put on a small index ***ee***, and above that a handle ***mm***, to turn the screw round as often as there shall be occasion, without at all endangering the displacing of the index; it being put on very stiff on a cylindrical part of the head, and the handle on a square. The screw has that third part of it which is next the plate larger than the other two thirds

[1] i.e. *carefully* [DS].

[2] i.e. *outside* [DS].

D. Sellers, *In Search of William Gascoigne: Seventeenth Century Astronomer*,
Astrophysics and Space Science Library 390, DOI 10.1007/978-1-4614-4097-0_17,

of it, by at least as much as the depth of the small screw made on it: the thread of the screw of the larger third is as small again as that of the screw of the other two thirds. To the thicker screw is adapted a socket *f*, fastened to a long bar or bolt ***gg***, on which is fastened the moveable sight ***h***, so that every turn of the screw moves the sight ***h*** either a thread nearer or a thread farther off from the fixed sight ***i***. The bar ***gg*** is made exactly equal, and fitted into two small staples ***kk***, which will not admit of any shaking. There are 60 of these threads, and answerable thereto are made 60 divisions on the edge of the bolt or ruler ***gg***; and a small index ***i***, fixed to the box ***aaa***, denotes how many threads the edges of the two sights ***h*** and ***i*** are distant; and the index ***ee*** shows, on the circular plate, what part of a revolution there is more; every revolution, as was said before, being divided into 100 parts. At the same time that the moveable sight ***h*** is moved forwards or backwards, or more threads of the coarser screw, is the plate ***pp***, in fig. 2, by the means of the socket ***q***, to which it is screwed, moved forward or backward, or more threads of the finer screw: so that this plate being fixed to the telescope by the screws ***rr*** in fig. 2, so as the middle between the sights may lie in the axis of the glass, however the screw be turned, the midst between the sights will always be in the axis, and the sights will equally either open from it or shut towards it.

Fig. 2 represents the moveable cover containing the screws to be fitly placed on Fig. 1; according to the taking off, as it were, or folding up of this cover, the inward contrivance of the screws and sights may appear.

And because it is conceived by some ingenious men, that it will be more convenient, instead of the edges of the two sights ***h*** and ***i***, to employ two sights fitted with hairs, therefore is added fig. 3, representing the two sights ***r*** and ***s***, so fitted with threads ***t*** and ***u***, that they may be conveniently used instead of the solid edges of the sights ***h*** and ***i***.

The 4th figure represents how the screws are to be put on. The tube ***AD*** is divided into three lengths; of which ***BC*** is to lengthen or contract as the object requires; but ***AB*** is here added, that at ***A*** there may be put on such eye-glasses as shall be thought most convenient, and, to set them always at the distance most proper for them, indexes or pointers, which here are supposed to be at ***B***; which length alters also in respect of divers persons' eyes. ***E*** is a screw by which the great tube can be fixed so as by the help of the figures any smaller part of it can immediately be found, measuring only or knowing the divisions on ***BC***, the distance of the object glass from the pointers. ***F*** is the angular piece of wood that lies on the upper screw of the stand[3], which is represented by figure 5.

For a description of the uses of this ingeniously contrived and very curious engine, see No.25, p.161 [*i.e. Philosophical transactions, v.25, 161*]

[3] This rest (by Mr. Hook's suggestion) may be rendered more convenient, if, instead of placing the screw horizontal, it be so contrived that it may be laid parallel to the equinoctial, or to the diurnal motion of the earth. For by that means the same thing may be performed by the single motion of one screw, which in the other way cannot be done but by the turning of both screws.

Chapter 18

Appendix 2 Selected Correspondence

Editorial Principles and Abbreviations

The extracts in this appendix have been selected with the following aims in mind: to gather together previously published correspondence of William Gascoigne; to present some of the newly discovered letters or transcriptions; and, above all, to illustrate the nature of the discussion that accompanied Gascoigne's pioneering techniques and his efforts to impart the theoretical basis of them to the satisfaction of his kindred spirits.

The extracts are presented in chronological order. The dates of the letters (as those in the rest of this book) are the Julian dates then in use in England. At the time, the year was usually regarded as beginning on 25th March and therefore the dates between 1st January and 24th March, say in 1641, were sometimes cited as 1640 or 1640/41. For the sake of avoiding ambiguity, such dates are given here as 1641.

A 'sources' section is included for all known letters—even those which have not been selected. This explains which of the sources has been relied upon for the transcription.

Standard contractions have been expanded for ease of reading: thus y^e, y^t, w^{ch}, etc., have been rendered as the, that, which, etc. Suffixes have expanded (such as -con to -tion and -õ to -on). Superscripts d and gr have been given as ° (degrees). The seventeenth century use of 'then', instead of the modern 'than', has been left unaltered.

Erratic punctuation (or the lack of it), for the main part, has been accurately reproduced—even though this sometimes makes for difficult reading. Very occasionally, however, full stops have been added where their absence could lead to misunderstandings.

Astronomical or astrological symbols have been converted into words (nouns or adjectives, depending on the context).

The reader should bear in mind that errors in the wording of the non-holograph letters may sometimes have been introduced by the original transcriber or copyist.

D. Sellers, *In Search of William Gascoigne: Seventeenth Century Astronomer*, Astrophysics and Space Science Library 390, DOI 10.1007/978-1-4614-4097-0_18,

There are only three holographic letters: the two from Gascoigne to Oughtred and the one from Gascoigne to Digby.

Most of the transcriptions represent only a part of the original whole letters.

Editorial signs

<text> uncertain reading
<–––> illegible script
[text] words added or suggested

1.
WILLIAM CRABTREE to GASCOIGNE
7 August 1640

Sources:

Philosophical transactions, v.27 (1711–12), 280–9, Observations upon the Spots that have been upon the Sun, from the Year 1703 to 1711. With a Letter of Mr Crabtrie, in the Year 1640 upon the same Subject. By the Reverend Mr William Derham, F.R.S. [source used]
The National Archives, RGO 1/40, ff.9v-10r [two paragraphs only, plus notes by Flamsteed]

I writ also to Mr. Townley at that time my Opinion in brief of the Suns Spots, (which you conceive to be Stars) and it seems he, or Mr. *Kay,* writ to the same purpose to you, desiring your Opinion: Which you freely deliver; for which I cannot but commend you, and especially for prefering Reason before any Mans Authority. Yet give me leave *(pace tua Amice desideratissime)* to speak my mind likewise freely concerning these appearances. I do not value the Authority of *Galilaeus* (though reputed the greatest Speculative Mathematician in *Europe)* nor yet *Kepler* (though *Astronomorum facile princeps)* further than either Demonstrative, or the most probable Reasons confirm their Opinions. Nor will I stick to subscribe to the Man whosoever shall bring better Reasons for his Opinion. I must acknowledge you say more for the stellifying of these Solar Obscurities, than I have heard before; yet I conceive not sufficient, either demonstratively or probably to countermand those which *Galilaeus, Kepler,* and others have produced to the contrary; nor yet such as can be cleared from such Objections, as Reason, Demonstration, and Observation may lay against them. My Occasions will not admit a full Disquisition hereof at this time; yet something I would say for the present, the better to furnish you where to object when I see you; that, so by diligent Inquisition, the desired truth may (may we have that happiness) be better found out by us.

I have often observed these Spots; yet from *all* my Observations cannot find one Argument to prove them other than fading Bodies. But that they are no Stars, but unconstant (in regard of their Generation) and irregular Excrescences arising out of, or proceeding from the Suns Body, many things seem to me to make it more than probable.

For first, for their Form; they are seldom round, but of irregular Shapes, and, as I have often seen, one side, or end of the Spot more thin than the rest, like to a certain misty darkness, and by degrees thicker, grosser, and darker, nearer to the main body of the Spot; just as the Smoak of some pitchy Fire, which is in one part very gross, and in another more rare and thin, turning at last into meer Air; Or like a Cloud, Fog, or Mist, more thick, dark, and gross in the midst; and more thin, fluid, penetrable, and transparent towards the sides; which I suppose is not compatible with any of the Stars.

Secondly, for their Colour: The lightness thereof differenceth them from Stars, or Planets; they being never of such absolute darkness as I observed *Venus* the 24th of November last: Tho' I have seen spots sometimes little less than she, yet always of a far paler and whiter Colour, looking (at least in some Parts) like some thin dissipated substance.

Thirdly, for the manner of their appearance. I have seen many Spots, which in the middle of the Sun appear of a round, body, but coming towards the side of the Sun, appear long. Which (if you rightly consider it) is a demonstrative Argument that they are not Globes, as all the Planets and Stars are: For Globes always appear of one form (round) in every Position; but Exhalations, or such likc fluid Substances, extended to a broad flat form, like our Clouds, which being over our Heads, and so in their full breadth, appear large and broad; but driven with the Wind, till they turn one edge upon us, seem of a long shape. So these Solar-Clouds, being turned about the Sun, may in the middle shew their full breadth to us, and about both edges of the Sun, turn their edges to us: Which answereth to the appearance.

Fourthly, for their continuance. Some of these Spots, arising at the East-side of the Sun, vanish before they come to the midst of the Sun. Others appear first in the middle of the Sun, and vanish before they come to the Western Limb; and for the most part they vanish before they have made a full revolution about the Sun. Which argues them to be but thin, vanishing, fading Substances, not like the permanent bodies of the Stars.

But to take off these Reasons, you answer, That you conceive these Spots to be Stars moving regularly in their own Orbes, which are many, though none of greater extent than about 1/10 of the Sun's Semidiameter from its Circumference; and that the swifter Movers in the lower Orbes, overtaking the slower in the higher Orbes, cause an appearance. You seem therefore to think, that they being so thin bodies, the Suns Rayes pass through them, and so one cannot be seen alone, till more being together, one heaped behind another, they stop the light of the Suns Rayes, and so cause an appearance. This I conceive is you[r] meaning: Or else (as you seem to insinuate afterwards) that the Higher reflects the Suns Rayes strongly enough upon the Lower (when they come within the Angle of Reflection) to make the interjacent Planet indiscernable.

But to these I answer,

If it be by their coming within the Angle of Reflection, that the light of the Sun reflected from the outer Planet upon the inner, doth make it (as you speak) indiscernable, then that Light so reflected is reflected either upon all places, as the Moons and Planets Light; or but upon one, as is the Reflection of a plain Looking-Glass. If the first, there would never be many seen (seldom above one or two) because the outermost would continually make the inner undiscernable. But *Gassendus* affirms, there are seen sometimes 40 at once in the Suns body. If the 2d, there would always be many seen, because the reflected Light would but occupy a little room, and that but for a small time, till the swifter were past the place of Reflection: Whereas many Days there are none at all seen in the Suns Hemisphere: And in both these cases, the outermost Planet of all would always in the space of 27 Days, be seen in the same place, being never obscured, none of the inferior being able to reflect Light upon it. Add hereunto, if any kind of Reflection should make them to appear bright like the Sun, and so not distinguishable from the Light of the Sun, what should[1] hinder, but

[1] NB (inserted by Derham): *Mr* Gascoigne *having, against these Words, inserted a rough-drawn Figure in the margin of Mr* Crabtrie's *Letter, I have also represented it in* Tab.2. Fig.4. *imagining it may somewhat explain Mr* Gascoigne's *Hypothesis, and what Mr.* Crabtrie *saith against it*

we should see them also bright Bodies by the side of the Sun, when they are passing either by the West, or East-side of the Suns Body? The Light being then reflected upon them by the inferior Planets as well as at other times, and that also upon much of that side of them which we should behold.

But if you wave this conceit, as insufficient, and fly to your former, That the swifter Movers in the lower Orbes, overtaking the slower in the higher Orbes, cause an appearance. To this I answer. 1. The thing you suppose seems to me neither necessary nor probable, nor do I conceive why they should not be seen, being themselves alone, as well as conjoined, seeing all other Stars and Planets are so. 2. If it be because they are of a thin, transparent Substance, till many, being one behind another, make them to seem grosser; Then they are not of the nature of other Planets, as is proved in Mercury and Venus, who of themselves appear dark Bodies, when they come between us and the Sun; nay, they must be more thin than our Clouds, which will easily be seen between us and the Sun, and hides it from us. 3. If it be because they are so little, that the Imperfection of our Glasses cannot discover one alone, there must be, without doubt, many Millions of them; which how they can be included within the compass of 1/10 of the solar Semidiameter, we shall consider anon. I have seen one of an ordinary darkness, (yea darker than many greater) yet not above 5″ Diameter. If this consist of two, or many, of themselves invisible, how many were in those which *Gassendus* saw of 1′½ Diameter? 4. The Figure of these great ones (being necessarily composed of Stars of such different Orbes and Motions) would quickly vary, by reason of the diversity of their Motions; like as we see in a Flock of small Birds. But 5thly, you say the furthest of these Orbes is not above 1/10 of the Suns Semidiameter from its Circumference. But there would not, in that small space, be room enough for so many Orbes of Planets, as have been seen at once. Which I prove thus. 1. *Gassendus* affirms there are sometimes some of about the 1/10 part of the solar Semidiameter; which is the whole space allowed by you for them all. And I my self have seen of 1/15 of the solar Semidiameter: And yet you must confess these great ones could only be the Conjunctions of some, not all. 2. There are many times seen in the Sun's Superficies, a great number of Spots, whose Diameters added together, would do more than twice fill the space you speak of. I my self have seen it, and so I believe have you. *Gassendus* affirms, there are sometimes 40 seen at once: If this was by Conjunction of Planets, in every Appearance, there was at least 8o Bodies at once on this side the Sun; it may be as many on the other side, besides those unseen (by your Reflection or otherwise) which doubtless must be far more than seen. For it is a most rare, and I think unheard of thing to see but 3 (which is less than the half) of our Planets, conjoin'd in visible conjunction at once: So that without question, if they be Planets, they are many hundreds; which must have so many several Orbes, and which certainly cannot be done in so narrow a compass, as the 1/10 of the solar Semidiameter. And that they cannot have any larger (I suppose not so large an) extent from the Sun's Superficies, may be proved by their motion through the visible Hemisphere of the Suns Spherical Body, by comparing the swiftness of their motion towards the middle and sides together. 6. If one of these (imagined) Planets be swifter than another, as they must needs be, then the conjunction of 2 or 3 swifter ones would make a Spot of speedier motion than the conjunction of 2 slower ones: But the motion of all about the Sun's Center, is always equal; yea, and the Spots retain the same Position one to another, (considering the Suns Sphericity, and the Angle of their appearance to us) just like the Fixed-Stars. So affirms *Gassendus, Moveri omnes eodem & uniformi motu, adeo ut, cum plures fuerint, nulla antevertat aliam, sed eundem tenorem in disco [Sun] perinde servent inter se, ac servant Fixae in firmamento.*

As for that other annual Motion of the Spots, you speak of, from West to East, upon their Axis inclined above 8 Degrees to the Ecliptick; I suppose it is not any real Motion of the Orbes of those Solar Planets or Spots, but only a visible Motion so appearing, caused (in *Kepler's* Systeme) by the Suns rolling upon its own Center in the midst of all the Orbes, not exactly in the way of the *Temporary* Ecliptick, but in the *Via regia* (as *Kepler* calls it) inclined certain Degrees to the *Temporary*; thereby turning about with him, the same way, his Adventitious, or Excrementitious Parts, the Spots, by his *Magnetical* or *Sympathetical Rayes.* And hence may be demonstrated the appearance of that Annual Motion in the Suns Spots you speak of. See *Galilaeus, Syst.Cosm p.339 & seq.* So also in *Ptolemie's* and *Tycho's* Systeme, the same Appearance may be demonstrated, supposing the Earth fixed in the middle of the Universe, and the Sun rolling round upon the same Poles of that *Via regia* (or way of the Spots) and keeping his Axis in Parallelism continually towards one and the same Part of the Universe. This may be certainly demonstrated, altho' *Galilaeus* there affirms the contrary. Other *Hypotheses* of that Motion may be feigned, as by the annual conversion of the Poles of the *Via regia* about the Poles of the Ecliptick in the Suns Body: But none I conceive so compendious, as the one of the former. For my part, I incline to the first: Yet if when we see you, you shew us any more likely Theory, for my part I shall be ready to consent to you in any thing with reason.

Thus you have, what for the present, I conceive of these *Maculae Solares. Fromundus* mentions one *Jo. Tarde Gallus*, who thinks them to be Secondary Planets; who hath written a Book of that Subject, and calls them *Astra Borbonia*: But I could never yet see it. What you, or he, or others may alledge for that Opinion, I know not. In the mean time it were too much levity in me, against my Judgment, to acknowledge them Stars; unless I see at least some possibility how they may be so, or some probability why they should not rather be Spots. Which when you, or they do produce from better grounded Reasons, Optical Experiments, or Demonstrations, I shall willingly recant my Opinion.

In the mean time, let me encourage you to proceed in your noble Optical Speculations. I do believe there are as rare Inventions as *Galilaeus Telescope*, yet undiscover'd. My living in a place void of apt Materials for that purpose, makes me almost Ignorant in those Secrets; only what I have from Reason, or the reading of *Kepler's Astron. Opt.* and *Galilaeus.* If you impart unto us any of your Optical Secrets, we shall be thankful, and obliged to you, and ready to requite you in any thing we can.

It is true which you say, That I found *Venus* Diameter much less than any Theory extant made it. *Kepler* came nearest, yet makes her Diameter 5 times too much. *Tycho*, *Lansberge*, and the Ancients, about 10 times greater than it was. So also they differ in time of the conjunction as far from the truth. By *Lansberg* the conjunction should have been $16^h\ 31'$ before we observ'd it: By *Tycho* and *Longomontane* $1^d\ 8^h\ 25'$ before. By *Kepler* (who is still nearest the truth) $9^h\ 46'$ before. So that had not our own Observations, and Study, taught us a better Theory than any of these, we had never attended at that time for that *rare Spectacle*. You shall have the Observation of it, when we see you. The Clouds depriv'd me of part of the Observation, but my Friend and second Self Mr. *Jeremiah Horrox,* being near *Preston,* observed it clearly from the time of its coming into the Sun, till the Sun's setting; and both our Observations agreed, both in the Time and Diameter, most precisely. If I can, I will bring him along with Mr. *Towneley* and my self, to see *Yorkshire*, and you. You shall also then have my Observation of the Sun's last Eclipse here in *Broughton,* Mr. *Horrox's* between *Liverpoole* and *Preston,* and Mr. *Foster's* at *London*. *Lansberg* in Eclipses, especially of the Moon, comes often nearer the truth than *Kepler,* yet it is by packing together Errors; his Diameters of the Sun and Moon being false, and his variation of the Shadow being quite repugnant to Geometrical Demonstration. His circular Hypotheses Mr. *Horrox* (before I could perswade him) assayed a long time with indefatigable Pains, and Study, to correct, and amend; changing and turning them every way (still amazed and amused with those lofty Titles of Perpetuity and Perfection, so impudently impos'd upon them) until we found, by comparing Observations in several places of the Orbes, that his Hypotheses would never agree with the Heavens for all times, as he confidently boasts; no, nor scarce for any one whole Year together, alter the equal Motion, Prosthaphaereses, and Excentricity howsoever you will.

Kepler's Elliptick is undoubtedly the way which the Planets describe in their Motions: And if you have read his *Comment. de motu martis*, and his *Epit. Astron. Copern.* I doubt not you will say his Theory is the most rational, demonstrative, harmonious, simple, and natural that is yet thought of, (or I suppose can be;) all those superfluous Fictions being rejected by him, which others are forced so absurdly to introduce. And although in some respects his Tables be deficient, yet being once corrected by due Observations, they hold true in the rest: Which is that argument of Truth, which *Lansberge's* and all others want.

Your conceit of turning the Circle into 100,000,000 Parts, were an excellent one, if it had been set on foot, when Astronomy was first invented. Mr. *Horrox* and I have often conferred about it. But in respect that all Astronomy is already in a quite different form, and the tediousness of reducing the Tables of Sines, Tangents, and all other things we should have occasion to use, into that form; as also some Inconveniences which we foresaw would follow in the composing the Tables of Celestial Motions, together with the greatness of the Innovation, deterred us from the conceit. Only we intend to use the Centesmes or Millesmes of Degrees, because of the ease in Calculation. I have turned the *Rudolphine* Tables into Degrees and Millesmes, and altered them into a far more concise, ready, and easy form, than they are done by *Kepler.* My Occasions force me to put an abrupt End to my unpolish'd Lines, and without more Compliments, to tell you plainly, but sincerely, I am

Your Loving Friend,
(*though* de facie ignotus)
WILLIAM CRABTREE

From my House in,
Broughton *near*
Manchester, *this*
7. August 1640.

Notes:

By W Derham: The Superscription of this Letter is, *To his Loving Friend Mr.* William Gascoigne, *at his Fathers House in or near* Leeds *in* Yorkshire.

The beginning of the Letter hath been torn off; but I find by that part of it that is left, it was Mr. Crabtrie's first Letter to Mr.Gascoigne, and that the torn part was only Compliments for his Writing to him, being a Stranger, &c.

By J.Flamsteed (1): This correspondence betwixt these two <ex…ious> arteistes was begun as it semes about the yeare 1639 or 40, upon occasion of the solar spots which, appearing frequently at that time, Mr Gascoigne, with Tarde, esteemed planets; Mr Crabtree solar or etheriall excressiencies. Mr Gascoigne esteemed them to move in orbes of which the remotest was not more than 1/10 of the Suns ½ diameter distant from his superficies, which Mr Crabtree esteemed scarce possible.

By J.Flamsteed (2): this first letter conteins severall cogent reasons to prove the solar spots no planets, orderly rankt & methodicall; which since they are plaine easie & I believe obvious in such authors as have wrote of those phainomena I omit. It was dated August 7: 1640: But I may not omit that in this Mr Crabtree affirmes the perpetuall tables of Lansberg & his hypotheses to be altogeather different from the heavens, & that hee affirmes that in the transit of Venus sub [Sun] hee found her diameter much less than any theorie extant made it—that Mr Horrox neare Preston observed it nearly from the time of its comeing under the Sun till the suns setting, & both our observations (sayth hee) agreed both in the time & diameter most precisely

Nor may I pretermitt this following paragraph [about dividing the circle into 100,000 parts] in this same letter since it shews the loftie & acute fancies of the Lancashire Astronomers.

2.
WILLIAM CRABTREE to GASCOIGNE
30 October 1640

Sources:

Philosophical transactions, v.30 (1717–19), 606–8. Extracts from Mr Gascoigne's and Mr Crabtrie's Letters, proving Mr Gascoigne to have been the Inventor of the Telescopick Sights of Mathematical Instruments, and not the French. By W.Derham, Prebend of Windsor, and R.Soc.Soc. [source used]

The National Archives, RGO 1/40, ff.10r-10v [two sentences only]

Something I am sure you were telling me concerning a way of observing the Places of the Planets by your Glasses. But I not a little lamented that my Time cut me so short, when I was with you, that I could not more fully ruminate and digest those strange Inventions which you shewed me, and told me of. My Lassitude after an unexpected and unacquainted Journey; my unpreparedness for those Cogitations (not intending that Journey the Day before) and the Multiplicity and Variety of the Novelties you shewed me, so wholly distracted my Thoughts into Admiration, that I cannot now give my Meditations any reasonable Account of what I saw: but must intreat you, in a few Lines, to rub up my Memory, and tell me again what you shewed me, and the Extent of those your Inventions. Which I desire, that I might consider, and rejoice to consider, how much and wherein Urania's Structure will grow to Perfection by your Assistance: and that (what in me lies) I may help you to remember when and wherein your Inventions and Observations will be of most use. I should also desire you to inform me what Bigness of a Quadrant you conceive to be large enough for Observation with your Devices. For I am e're long going to Wigan, 12 Miles from hence, where much Brass is cast; and then I could see whether I could procure such an one cast. You told me (as I remember) you doubted not in time to be able to make Observations to Seconds. I cannot but admire it and yet, by what I saw, believe it: but long to have some farther Hints of your Conceit for that Purpose. One Means, I think, you told me was, by a single Glass in a Cane, upon the Index of your Sextant, by which (as I remember) you find the exact Point of the Sun's Rays. But the way how, I have quite forgotten, and much desire. Your Device for the exact Division of a Quadrant, by dividing 11 Degrees into 10 Parts, I did then understand, but do not now fully remember. If it might not be too much Trouble to you, I should intreat you to give me such a Paper-Demonstration thereof as you shewed me, and two or three Lines plainly of the Use thereof, how to find those small Parts. I lost the little Paper, wherein I noted the Moon's Diameter, which we observed when I was with you: I pray you send it me, if, &c.

I cannot conceal how much I am transported beyond my self with the Remembrance (of that little I do remember) of those admirable Inventions which you shewed me when I was with you. I should not have believed the World could have afforded such exquisite Rarities, and I know not how to stint my longing Desires, without some further Taste of these selected Dainties. Happier had I been, had I never known there had been such Secrets, than to know no more, but only that there are such. Of all Desires the Desire of Knowledge is most vehement, most impatient: and of all kinds of Knowledge, this of the Mathematicks affects the Mind with most intense Agitations. I doubt not but you can experimentally witness the Truth hereof, and one time or other have been no Stranger to such Thoughts as mine. And therefore although Modesty would forbid me to request any thing (until you give me leave) but what you please voluntarily to impart, yet the Vehemence of my Desire forceth me to let you know how much I desire, and how highly I should prize any thing that you should be pleased to communicate to me in those Optick Practices. Could I purchase it with Travel, or procure it with Gold, I would not long be without a Telescope for observing small Angles in the Heavens; nor want the Use of your other Device of a Glass in a Cane upon the moveable Ruler of your Sextant (as I remember) for helping to the exact Point of the Sun's Rays. But seeing Urania is, &c.

Notes:

By J.Flamsteed: To this first letter [i.e. 1640 Aug 7] of Mr Crabtrees I find <no> answer. but by his second letter dated October 30th 1640 it appears that betwixt those dates Mr Crabtree visited Mr Gascoigne with some friends at Middleton where hee viewed all Mr Gascoignes new devised instruments & measured with him the moones diameter <—> but at what time the observation was made & what it gave is not mentioned, the note being lost in which it was written: After very ceremonious thankes for Mr Gascoignes courtesies & entertainments, in the preface of this letter Mr Crabtree complaines that hee had forgot some things what hee had seene and intended to write to him: then falls on the businesse of the solar spotts of which haveing made, saies hee,

> some observations formerly; I could never observe any difference of their situations one to another, but that of their spreadeing wider asunder towards the middle of the sun, & nearer togeather toward the sides (which they must needs doe, by reason of the Sun's spaericitie, if they be contiguous or neare the suns convex superficies & turned about with it.

After this hee proceedes to demonstrate that the Sun's spotts are not above 1/10 of his semidiameter from his superficies, which hee does by computeing how far they, moveing 60 degrees in such an orbe, would be removed from the suns center, & compaireing the motion so found with what hee has observed, hee allowes them one apparent revolution in 27 days, so that they may move 60 degrees in 4 days, 12 hours, which considering them to move in an orbe 1/10 of the Sun's semidiameter, from his superficies, will cause that (the suns semidiameter being supposed 90 ptes) a spot having moved 60 from his center should appear only 3 4/10 such partes from his limbe. But saies hee,

> I have often observed the motion of divers spotts, & could never find any gone from the middle of the sun so far towards his side in 4 dayes, 12 houres.

Afterwards he demonstrates that the illumination of the spots by the sun lessens not their diameters more than 1/3 of them, but that they are seene far more contracted by reason of their contiguity to the sun: with some other things materiall to his purpose, but not to my designe in this transcription which therefore I omitte.

Next hee argues Lansberges diameters of the earths shadow of falshood from his owne fundamentalls but of this I suppose Mr Horrox has enough in his booke against Lansberge, whose absurdities and causlesse boasts are well noted both by Ricciolus in his Almagestum Novum & others which allso therefore I pretermitte.

Hee requests at the last an observation of the distances of the Pleiades to which hee had observed some appulses of the moone then notes severall opportunitys for observations of the Moon passing neare fixed starres then with some necessary directions how to continue the intercourse of theire letters safely, advertisements concerneing some bookes of Keplers & some other small request for himselfe & a friendly encomium to his correspondent concludes his letter.

3.
GASCOIGNE to KENELM DIGBY
30 November [1640]

Sources:

British Library,MS.Add. 61873, North (Sheffield Park) Papers, xiv, ff.54–55

To the right worth his noble friende
Sr Kenelme Digbie Knight
this
present
S^{r}.

I have herein enclosed certeine of the choisest scheames[2] apperteining to perspectives, with facile & briefe precepts derived from as few necessarily knowne lines as are possible: In the 1 diagram are required WB½ the axis of the sphere (in this supposing vulgar glasses) wherof akHS is a segment: Bi½ the greatest widenesse admitting AB a paralell ray of incidence at B refracted unto C the apparent concourse of all paralell incidences upon B supposed to be revolved upon i. WB, iB, iC will command all or any of the rest being aided by Sk the glasses thicknesse of much use when we have to do with oblique rayes of incidence as Zm. My desire to be understood pre<vayl>s with me though I know you litle neede it to adde one line of explication: WB, BÏ (with WiB) gives Wi w^c + ik = Wk · ik + kC=iC. <iC&> iB with BiC gives iBC w^c less iBg = &c (I use this · as a sign of coniunction & seperation (to avoid repetition) <where> + or—immediately precedes or followeth it. Those observations of the Sun & Starres here denoted are not by me beleeved to be true <to> a second though I am confident of the possibility of that by a glasse no longer then that you did see at Yorke if the scale, case, & glasses were carefully made by skillfull artists. When I have the last according to my owne desire I can readily procure the other; in the interim I am content with some apparent error which may easily be deduced from ac, dc, bc in the Pleiades wherein I might if I would with as much facility have inserted 30 other starrs exactly (if I here entended, not more, then to shew what may be don) in their naturall scituation; for I finde it possible to set the brasse pointers that they will shew onely part of 2 of the least starres visible in a perspective:

And I have an other additament that will afforde the inclination of one star to an other in any verticall of the excellent use for the invention of their longitudes: How far Morinus would have stretched such an invention for his owne, not Countries profit had he knowne it his book witnesseth; Contrary to whom I also hope to evince the Moon's distance from the Earth onely from its spots: Each dayes alteration of its diameter is as conspicuous as the encrease or diminution of its light. Neyther are the other planets priviledged by their very acute angles, nor need Galileus proiect of a hill & Collumne or Cilinder looked at at divers miles distance; but may by a large & good perspective glasse have their diameters defined in some <par>ts of seconds or decimalls answerable as I have experiment<ted> <on> Mars by a glasse of 5 yards length which I have not yet fitted for cu<—>s or ready use. This draught of the ½ scale wi<ll> serve any in<dust>rious artisan to draw a perfect one by: All <—> parts of it by the directions I have sent & a litle practise will <be> easily appliable to their proper use. It must be placed in the place of concour<se> in the case betwene bothe the convexes: & <it>s distance from <the> outmost glasse measured in all uses by the same parts that the fixed part of the scale is divided by. If in any part I be not understood upon your information of me I shall be ready to render it more at large. If at any time you please to write a letter left eyther at the Beare in

[2] The 'scheames', or diagrams, that originally accompanied this letter seem to be missing The labelling and description, however, seems to match the diagrams that accompanied Gascoigne's letter to Oughtred about three months later (see letter 8).

Basing Hall Street or sent to the Post that comes to York will not faile of being delivered if directed thither whence this is dated.

If in a darke house you receive severall species through a convex glasse (after the usuall maner) upon a white paper scored with black lines & interpose a very convex glasse betwene your ey & the paper it will be easy to finde a position of the paper wherin the lines will appeare (being not at all reflected) next you and the other species or pictures set of at severall distances, very probabely according to the strength of the motion of the rayes producing them: If you observe this it may come to passe that the strength of your judgement may force an unanswerable argument from it. If any mathematicall novelties come to your notice (which no rare book can escape) that I may be onely informed of their worth & subiect shall no litle augment that obligation wherby I am bound to be

Noble S^{r}
your observer
William Gascoigne
Middleton neare Leedes
Noveb: 30

4.
GASCOIGNE to WILLIAM CRABTREE
2 December 1640

Sources:

The National Archives, RGO 1/40, ff.10v–11v
Bodleian Library, MS.Eng.7031, ff.43r–44v [source used—part only]

Mr Towneley thinkes he can procure me a peice of Iron to my mind to forme glasses on as I think best, which if I get I doubt not in a short time to be furnished with glasses of speciall power. I have sent you the prime diagrams appertaining to prospectives (as the glasses are usually made) with facile & breefe precepts derived for as few necessarily knowne lines as may be: *WB* ½ the axis of the spheare whereof *akFS* is (diag:1) a segment *Bi* ½ the greatest widenesse admiting *AB* a parrellell ray of incidence at *B* refracted unto *C* the apparent concourse of all parellell incidences upon *B* supposed to be revolved upon *i* . *WB* . *iB*. *iC* will command any of the rest being aided by *sk* the glasse thicknesse of much use when we have to doe with oblique rayes of Incidence as *zm*. My desire to be understood prevailes with me (though I beleeve you litle need it) to add one line of explication. *WB*, *Bi* (with *WiB*) gives *Wi* which more *ik* is equall to *Wk*. $ik + kC = iC$. *iC* & *iB* (with *BiC*) gives *iBC* which lesse $iBg =$ &c. I use this as a signe of conjunction & separation (to avoid repetition) where + or—eyther immediately anteceds or succeds. The observations of the Sun will manifest unto you the requisites, onely where *dh* stands in the 2 and 4 diags: I set a crosse like this [fig. 18.1] covered with paper on the barrs drawne with circles from the center the rest of [fig. 18.2] the Sun's rayes that goe to the other table at *GE* give its breadth there. the square props resible its place for the ruler to goe in, by this at one observation I have what is required in the diagram.

Fig. 18.1

Fig. 18.2

I dare not affirme these observations very true, onely here intending to declare how wee may make true ones; I shall have shortly another way whereon I most rely, In the scale drawne on the Cards you must suppose the moveable *gesb* let into the fixed *wrqd* (all the black barrs *w*, *x* &c, & all that at <z> being soe cut out) that they may make one plaine.

WB the radius of the glasses convexitie *Bi* the ½ widenesse of the glasse through which the Sun's ray *AB* passeth to *C*.

Notes:

The Bodleian manuscript does not record who wrote this letter and gives no date. Fortunately, at least 13 lines of this letter are also in the letter of 2 December 1640 from Gascoigne that has been transcribed by Flamsteed.

The letter terminates with detailed calculations and observations (omitted here), which are very similar to those appended to Letter 8 (the second letter to Oughtred)

5.
GASCOIGNE to WILLIAM OUGHTRED
2 December 1640

Sources:

Cambridge University Library, Portsmouth and Macclesfield Collection, Add.9597/13/5/103,r & Add.9597/13/5/103a,v. [source used]

Rigaud, S.J., Correspondence of scientific men in the seventeenth century (1841), pp.33–4

S^{r},

amongst the mathematicall rarities these times have afforded there are none of that small number I (a late intruder into these studies) have yet viewed which so fully demonstrates their Authors great abilities as your Clavis, not richer in augmentations, then valuable for contraction; that forceth me to censure my self presumptuous in this action; & this tedious, that I have not allready related the occasion. The choisest forreigne wits within these 30 years as Galileus, Scheiner, Keplar, Hortensius, & Monsieur de Cartas, have so diligently pried into perspectives that it may credibly be thought, that their rarest use is allready known, yet so it is I have eyther found out, or stumbled on, which indede they have allwayes stumbled at, a most certeine & easy way, wherby the distance betwene any the least stars, visible onely by a perspective glasse, may be readily given, I suppose to a second; affording the diminutions and augmentations of the planets strangely precise, as also their centers; stretchable to the invention of the moones true parallax of altitude by its owne body, & of the inclination of one star to another of a well knowne site; & able to bring sufficient aid to your aged eyes to finde all requisits (according to yours of navigation) at a large distance for searching the Earths diameter. It is a novelty capable of such frequent use, that before it travell to other able judges, may I receive that favour it shall undergo your experiment & censure. I bestow onely part of that time in these studies, which other Gentlemen, our neyghbors, spend in hunting, & want that great helpe an able Artist might afford, And beleeve that you will finde those errors I see not. If you please to grant my request I may be readily enformed by a letter sent by the northern Post, or left at the Beare in Basing Hall Street with the Wakefeild Cariers; directed to my Fathers house in Middleton, neare Leedes, where remaines,

S^{r},
Your troublesome unknown friend,
Will. Gascoigne
Decemb. 2, 1640.

6.
WILLIAM CRABTREE to GASCOIGNE
28 December 1640

Sources:

Philosophical transactions, v.30 (1717–19), pp.608. Extracts from Mr Gascoigne's and Mr Crabtrie's Letters, proving Mr Gascoigne to have been the Inventor of the Telescopick Sights of Mathematical Instruments, and not the French. By W.Derham, Prebend of Windsor, and R.Soc.Soc. [source used]
The National Archives, RGO 1/40, f.12v

My Friend Mr.Horrox professeth, that little Touch which I gave him of your Inventions, hath ravished his Mind quite from it self, and left him in an Extasie between Admiration and Amazement. I beseech you, Sir, slack not your Intentions for the perfecting of your begun Wonders. We travel with Desire till we hear of your full Delivery. You have our Votes, our Hearts, and our Hands should not be wanting, if we could further you.

Your Diagrams for Perspectives I have viewed again and again, and cannot sufficiently admire your indefatigable Industry, and profound Ingenuity therein. I am much affected with the Symbolical Expressions of your Demonstrations. I never used them before (but I will do) yet I understand them all at the first Sight, and see well the Truth of your Demonstrations.

Notes:

By J.Flamsteed: Mr Crabtrees answer to the letter [Letter no.4 here, 2 Dec 1640] and paper I find dated Dec 28. 1640 with a postscript Jan. 5 40/41 in which the argumt concerneing the solar spots is continued. whose true motion about the sun hee concludes to be in 25 dayes tho the apparent by reason of the earths motion be in 27 dayes. in this letter hee treats at large of the best method for observeing them & deduceing theire places & the angle of their visible way or their via Regia (as he stiles it) with the Ecliptick afterwards expressing his doubts concerneing the quantity of the moones semidiameters & the distances of the Pleiades he writes thus –

> The moones semidiameters you say, you believe all save the 2 last should be lesse by 26″ or 28″: and yet in the Catalogue of observations you set downe thus semid moon 16 No: 15–14 ….. obser. 26″+. and after thus 15′–40″ by the tried glasse so that it seemes the tried glasse made the observation more not lesse. (except by the tried glasse you

meane the <misplace> glasse) & therefore I know not whether observation to take for truth Nor whether the distances of the Pleiades in the table ab 35′–01″, ae 27′–50″ or those in the[3] letter ab 35–32. ae 28–18 I pray therefore give mee a word of directions in those & whether you conceive the semidiameter of 16′–10″, 16′–11″ need any correction because of any error in the glasse.

at last requesteing an explication of some occurring difficulties in the 5 opticall schemes[4] hee concludes

For this last doubt I find an explication in Mr Gascoignes next letter which tho a little out of order may most fitly bee here transcribd. Tis thus [Letter no.7 Jan 25, 1641/40]

7.
GASCOIGNE to WILLIAM CRABTREE
25 January 1641

Sources:

Philosophical transactions, v.30 (1717–19), pp.604. Extracts from Mr Gascoigne's and Mr Crabtrie's Letters, proving Mr Gascoigne to have been the Inventor of the Telescopick Sights of Mathematical Instruments, and not the French. By W.Derham, Prebend of Windsor, and R.Soc.Soc.
The National Archives, RGO 1/40, ff.12v–13v
Bodleian Library, MS.Eng.7031, ff.35r–39r [source used]

S[r]

I have soe accustomed my selfe to observe by direct contraries, that I cannot in this contradict custome, nor beleeve my selfe immethodicall though I begin with your conclusion; a straight course would not sute with refractions, which keepe such unknowne pathes that the excellent Kepler professeth in pag 113 Paral in Witellion: In genuina huius rei causa directe & a priori demonstranda haereo[5] & for feare it should passe unnoted himself thus Marginal notes it. Dic quibus in terris, et eris &c.[6] And in the next page save one he takes more paines to get trueth out of false hoods (vitelliors erronious or grosse observation) then would have manifested its elements, had he in this beene as industrious as in most other searches. I hope I shall in this soe clearely prove it that when you receive your booke I doubt not you will upon persusall concurr that in what concearnes this he is mistaken: It seemes he mised the Cushion there forsaking his trusty Masters to whom he thus testifieth his effection; in pag: 95 Plurimum namque amo analogias fidelissimos meos magistros, omnium naturae arcanorum conscios[7]. Though he in this neglected them yet in no other matter doe the[y] merit more attendance then here. himselfe in pagi 19: Paralip

[3] *Flamsteed notes: 'this letter lost'

[4] The text of this request is probably that given in Letter 27 (which is erroneously dated in the Bodleian Library manuscript).

[5] "In demonstrating the true cause of this directly and a priori, I am stuck" –Johannes Kepler, *Optics: Paralipomena to Witelo & Optical Part of Astronomy*, trans. Donahue, W. (Santa Fe, 2000), 126

[6] Dic quibus in terries, et eris mihi magnus Apollonius. According to Donahue (*Op.Cit.*, 126) « tell in which lands, and for me you will be the great Apollonius [Apollo] », a parody of Vergil, *Eclogue 3*, 104

[7] "I love analogies most of all: they are my most faithful teachers, aware of all the hidden secrets of nature" – W.Donahue, *Op.Cit.*, 109

Table 18.1

		°	′			
The Angles *feg* in the water	45	71	40	Prop:30°:42.15′::35	50	28
	40	59	45	45	71	58
	35	50	27	40	59	49
	30	42	15	25	34	38
	25	34	46	20	27	23
	20	27	23	15	20	22
	15	20	16	10	13	30
	10	13	30	5	6	44
	5	06	45			
		ced = *fak*				

& pag: 651: Astron: Coper: saith that supposeing in the 1st scheame[8] *ne* the handle of a pair of ballance, *a* one end of the ballance *g* the other, *e* the center or pin whereon they move, & from *a* one weight hanging, from *g* another, the proportion of the weight *a* to *g* is as *np* to *nf* & *p* & *f* are found in *no*, by *ap* & *gf*. Kepler testifieth in the page preceeding the last cited the same in this mannor: Praeterea omnis motus naturalis vel artificialis, in quem vel eadem vel analoga concurrunt principia, dispensatur per sinus angulorum: praecipue vero et euidentissime, motus vel nisus brachiorum in libra & statera[9]. Then if we suppose the weight at *g* to be let into water according to the proportion of the solides of the weight & water (suppose *nf* to *nh*) *g* will ascend here to *i* therefore *a* must needes descend to *b*=to *ns* found in the handle by *bs*: & *i*=*nhi* perpendicular to *ih* & as *a* to *i* so *np* to *nh*. Then to apply this to refractions let *neo* be perpendicular to the horizon. *gea* a ray in the free aire cutting *no* in *e* with an Angle of 30° (suppose as I used in observing this underwritten table [Table 18.1] a paire of well polished brass plate at *g*, *gf* the sine of 30°, *ef* of 60° of a sufficient length then placeing water that its superficies be *ql*, the eye placed at *c* will discerne *g* as if it were ascended into *k*, *feg* =*aeb* being 30° *ced*=*mek* will be 42°15′ (if the aire be to the water as when I made my observation) & the Angle of refraction *gek* may be in all conclusions precisely found being =*ced* – *gef*, the very same here as before *gei* the difference of the weight *g* in the aire & in the water which difference may there alwaies be found by *fg*, *hi*: here in like sort by *fg*, *mk*=*cd* which fully appeareth from this observation.

I take the proportion of 30° to 42:15′ because I could not by reason of the obscurity that the water made at 45° & 40° soe perfectly see the brasse plates at 71:40′. 59:45′ & becaus I let the water flow over *e* untill it made one plaine supeficies which

[8] The drawings are missing, but the labelling seems to be generally consistent with those accompanying the February 1641 letter to Oughtred (Letter 8 here)

[9] "Furthermore, every artificial or natural movement, in which the same or analogous principles concur, is measured out by the sines of the angles; but principally and most clearly, the movement or tendency of the arms in the balance and in the lever" - *Epitome of Copernican Astronomy*, CG Wallis translation, 130

might happily bring more inequality at the greater inclinations then the lesse: otherwise the greater & better. the Angles *ced* were taken by an instrument as carefully as I could & the proportion of the sine of 30° *fg* to the sine of 42° 15′ *icd* are soe neare in all that I suppose more exactnesse were needlesse, & it is by this apparant enough that *cd*, *fg* ar the true proportionators of the severall inclinations. this observation differs onely an unit from De Cartas 250 to 187 he makes the reason of refraction to be because *be* a ray of incidence moves more easily in the water then in the aire & therefore tends towards *eo* as *eg* whereas if a bullet were to move in aire from *b* to *i* & meet the water at *i* [*e*?] it would tend towards *k*. If *qel* be the edge of a peece of glasse pollished & *fg* another edge parrallell unto *ql*. the proportion of *cd* to *fg* is readily found by placing any thing at *g* that can be discearned through the breadth of the glasse or if *rbp* be a right Angle triangle (as in the 2 scheame) of the glasse pollished on the edges & a ray of the sun *ab* be admited in a dark roome through 2 smale holes to fall perpendicular upon *br* it will passe unrefracted to *b* by which it is turned to *I* then measureing on the ruler whereon the glasse with the 2 runers that have the 2 holes in them, are placed *IP*, which by the 2 leggs of the rectangle triangle (of glasse) *Pr*, *rb* will show *bIr*, which is to *bp* as *PI* to *PbI* the sine complement whereof is *bx=n<f>* & *al=rp* soe we shall have the 2 sines measuring all refractions in the peece of glasse however it be after formed.[10]

If in the 3rd scheame *fgh* be the triangle or prisma of glasse & *cb* the suns ray admited through the holes, it will be refracted from *b* to *m*, from *m* to *p* where it is received on the ruler parallell to *gh* or table parallel to *fh*, *ph*, *hm* (for you may marke *in* on the glasse with Ink or some thing else) give *mph*, which is to *lmn* as *ebc* to *kbd* + *lmn* = *ebc* = *gfh*. If *f* be placed on the ruler & its perpend. fall betweene *fg*, *fh* you will find in all inclinations *ebc* + *mph* equall, because *kbd* + *lmn* are also, alway is equall & that to *gfh*.

If in the 1st Diagram that I sent before you take *AB* to be *ab* in this 2 scheame, *ABW* there *abl* here & *BC* in that for *bI* in this. And the tang: of the Arch *kBH* at *B* (the refraction of each point in *a* being regulated by its tangent, which represents *qel* (in all inclinations) in the 1st scheame, where is the same as *b* in the 2: & *B* in your 1st diagram. for *BP* also *C* for *I* you will readily perceive that *GF*, *DE* are the true regulators of all change; And how to compare the 2nd part of the 1st diagram with the 3rd scheame: & soe in the rest.

Hence I proceed to the other requests in all onely labouring for to be fully understood. first I find as neare as may be (in the first Diagram) by 3 points *Bkb*, *WB* ½ the axis of the spheare whereof the convex glasse is a segment, by this semidiamiter I cut a peece of fine pastbord which I apply to the glasse, if it be too litle I increase my semidiamiters, if too great diminish, untill I come soe neare the truth that there be no difference betweene the convex of the glasse, & concave of the pastboard. thus I get WB (which labour were needlesse if we know the axis of the toole

[10] The labelling of the 2nd 'scheme' referred to here is not the same as that which was sent to Oughtred (see fig.18.6) and errors and inconsistencies in this copy letter make it difficult to infer what the 'scheme' was.

forming the glasse) then I measure ½ *Bb* the glasse breadth= 2*ib*. whence *ik* will at pleasure be produced: These knowne I place the glasse in its case on a ruler in the usuall sun observing fashion at its accustomed length, laying the concave glasse apart, & move the table whereon the sun is idolized betweene 2 and 4 inch (for that I beleeve will serve in the glasse) of the concave glasse holding end of the case until I resolve where it receives & represents the Sun most lively, there I screw it fast; & by the helpe of a strong wier which hath a small moveable thread tyed on it. I measure (puting through a hole purposely made a litle below the center of the table) *kC* exactly (*C* being the center of the Sun's representation) which + *ki* = *kC*. thus I find *DBE* according to the rules I formerly sent you. Because *Bi* hath no sensible proportion to the Sun's distance, I suppose all the Sun's rayes whereof *AB* is one to be perpendicular to *Bi* or parallell to *Wk* or more fully in the other ½, because *iv*. *vb* compared to *iz* the Sun's distance from *i* is insensible, therefore I suppose *omz* null: or *zmb* + *mbq* = *ABC*. I beleeve it possible by an excellent glasse wherein *iC* is 4 foote *iB* neare 3 inch to find by one observation the true place of any object as z within 1½ mile. It seems Schriner pag.99: refract[11] with his acquaintance Malapart were deceived concerning the alteration of the distance of the 2 glasses in the 4th diagram being lesse in Summer then in winter which the[y] thought the Sun's approach to the earth altered, which is proved by calculation impossible; the true reason being that aire as it is more or lesse condensed changeth the proportion of refraction of unchangable glasse, which alteration is more observable betweene aire & glasse, or glasse & water, then aire & water.

I thought I had somewhere in my last letter shewed how *icd* in the 3rd Diag: was said to equall *fge*+*geb*. If I take the last observation in the former paper wherein I perceive I am not understood for that Angle 14′ 57″ *qkl* (which you I perceive tooke for the Sun's semidiamiter) was the Angl both by day & night in the glasse that 1 yard made being placed for tryall of the glasse trueth 11½ score yards from the glasse, which makes the true Angle *fkp* 14′:56″ 1″ lesse then the observation (The note that was under) these 2 observations &c was set there because if *kl* 4362 were the true distance of the Sun's picture from the glasse, in the same glass this *kl* (*q*) of 11½ score will not be just 4533 as the calculation will prove) I find that *DE* in the 1st Diagram is to *GF* as sine 2. 5′.52″ to sine 1. 19′.56″. Then I say as 2. 5′.52″.1.19′.56″ :: *lkq* (or *icd*, suppose) 15′ to *gcb* 10′ -. Therefore as *bgc* 89. 50′ + to *gcb* 10 - :: *bc* 10½ (hundreth parts of an inch the glasse thicknesse) to *bg* 0.03 (lesse then the 300 part of the 100 part of an inch.) Therefore 3600 (1 yard) lesse *ea* = *gb* 0.03 is equall to 3599.97 + insensibly differing from the truth. As it is with *c* in the 3rd <1> soe with *k* in the 4 & 5. this last is the Angle measureing the glasse, where *q* is the same point that *C* is in the 1st diagram, if here you place the scale it measures *ql* or if here a haire be set that it appeare perfectly through the glasse *ab* you may use it in a quadrant for finding of the altitude of the least starr visible by the perspective wherein it is. If the night be soe darke that the haire, or the pointers of the scale be not to be seene, I place a candle in a lanthorne soe as it cast light

[11] Scheiner, Christoph: *Refractiones Coelestes* (1617)

sufficient into the glasse, which I find very helpfull when the Moon appeareth not, or it is not otherwaies light enough. The reason why the glasse *ab* in the 5 Diagram makes noe alteration is because as it multiplyeth the species coming from *k* it also multiplyes *ql* the remove of the scale from the unrefracted line *qk* so that if *ab* multiply *ql* that it seeme 4^{in} or a more convex *ab* augment it that it show equall to 8 or 12^{in} the number or parts in the scale will never alter if *kq* keepe the same, therefore it is the most simple exact & ready way that is possible for the measureing of the planets diamiters. And if in moveable scale the 2 parelell barrs betweene 2 & 3. 6 &7 soe made as [you] may move into the midst of the case as that betweene 4 & 5 that I sent you which here must be no longer then the 2 outmost & 2 plates of 20 a piece let into the case & soe fastened to the ingoing bars that one plate may exactly cover the other within the case, & there may be opened to measure any Angle betweene 20′ & 40′ without any damage to the eye, you may measure both the Sun's & full Moon's diamiters at pleasure: if *ab* be the 2 sides of the plate 20′ or some knowne parts of the scale neare it *cd* the 2 barres let into the case *e* the shaft where the screw is to be placed *df*, *gc* the gradiated part of the scale & both plates soe fastened to either ½ of the scale that *a* of one plate concurr with *b* of the other when both halfes of the scale are moved most inward it will be capable of covering the Sun, & by the distance of the *b* of the one from *b* of the other removed until they cover the horizontall diamiter of the Sun give the tangent of that Angle on the scale: I thinke you will understand what I meane by this Marginall rude draught.[12]

The proportion of the glasse *ab* is impossible to be truly had at London because it cannot be given in Circles, the nearest if the glasse have the same refraction as that above will be that the superficies of the next *m* in the 5 Diagram be a segment of a globe the axis whereof is 3.46; the axis of the other superficies next *k* 19.00. or as 2.3 to 13.2 such as the thicknesse thereof be 0.3 the true proportion if the[y] were made of ovales to unite all the arayes at *m* is greater. the breadth of *ab* you may proportion as you would make the greatest *qkl* either 1° or 2° so that the case be made wide enough, if it were more such is the advantage of this glasse that the eye neare *m* will admit all the rayes at once: yet because *ab* cannot be wider then the axis of the lesse spheare, the greater distance of *a* & *b* must have greater axis; which because the[y] are lesse augmenters then the segments of the lesse speares will also be more transparent with lesse widenesse of the glasse *k*, by which having pitched upon a widenesse of *k* for any knowne convex *ab* we may increase *pf* as we please: as if *k* be one inch wide, *qkl* = *pkf* 15′ which after the conversion at *m* is 6° as the Cube of the sine of 6° to that of the sine 15′ soe is the augmentation of the solid by *ab* above *ql*. suppose this augmentation be of sufficient perspicuity or void of spots, & by this I would find an augmentation of the solid of *pf* in a given number reserving the glasse *k* whereof I onely must increase the widenesse, but placing another *ab* of power proportionable to the demaund, as if it were required to double the former augmentation of the sine *pf* 15′ after *m* 6°. twice the former widenesse of *k* will be sufficient for such an *ab* as will make the Angle after *m* 12°: the proofe &

[12] No 'rude draught' was included with this copy.

1st notice of this I valew amongst the rarest of all I have yet light upon for the augmentation of species.

If you send to Mr Towneley to provide you any glasses, let *ab* be 2 inch wide soe it will measure 2°+ if the distance betwixt *k* & it be not above 50 inch & the burning point of *ab* some 3 inch, if he try not this in the Sun he will be deceived in the other glasse *k*: if you give directions for a wider *ab*, all other things answerable, the most usefull will be those that have the burning point in least remote. The glasse *k* cannot be too wide if you have an *ab* of 3 inch from *m* & of 2, & of 1, but to get this a[t] London is impossible except one had a case of purpose & by trying by them selves all that one could conveniently meet with, conditioning for such a price for each one keepes, accidentally one may stumble of 2 or 3 such as would indiferently agree: It would be needfull to have the axes of each, & a peece of glas of some convenient forme (whereof before) for the measureing the proportion of the refraction; & all these of one & the same glasse. I had a letter from Mr Towneley who cannot meet with Rifler the best glasse grinder, nor a man that can forge iron into any usefull toole for my way, which must have it void of flawes, I have given directions to a Trooper an excellent Gunsmith whose skill as soone as Ale will allow I shall see. It may be ere long I shall have occasion to visit London where I shall fit my selfe. If you come at any time neare Wigan inform your self as fully as may be how many pounds of brasse they can melt into any forme if they could melt brasse to cleave to iron as I have seene some great ordnance it would be much cheaper to have a quadrant cast at an iron furnace & after on one side covered with brasse, on the rimme thick enough to receiv a mult rimmed cursor, which you will find readily affording 10 or 100 more divisions then those 2 in Morinus, in those hastily writ directions, I forgot to tell that *ad* should be 10 or 20 centesmes if the radius were 4 or 2 foot in that of 4 foot a 6 rimed cursor will give each 2000 part of a degree, one of 26 rimes each 10000 part; a number beyond the exactnesse of any table of sines yet extant. It seems strange to me to find Lansbergs excentricity of the Moon so exactly bisected, & such great probabilities that the Sun goes the whole; before I come to eyther I will set you straight in what you doubt in my last letter as semid: of the Moon 16 November 15′:14″ .... observed lesse by the true glasse 26″ & more lesse thesse 26″+ belonging to the whole diameter, I calculated the semidiamiters in decimalls & find the observation (in the glasse tryed by the yard 11½ score distance by day & night) 0.0043 lesse in the Moon's semid: by the true glasse then the other therefore 31″ are to be substracted out of all the former observations if the formerly used glasse *ab* in the 5 Diag: removed not: *ab* of the Pleiades in the letter 35′, 32″ - 31″ = 35′, 1″ *ab* of the note which I set for the truer, & said that 26″ or 28″ might be substracted out of the whole Angles of the Pleiades in the letter because the Angle in the glasse at 11½ score distance was 2″ lesse in the true Angle then in the glasse as you may see in the observation where the ½ Angles of both are set downe therefore I set rather – then +. That night in the faulty glasse *kq* in the 5 Diagram was 4400, *ql* (the Moon's semid: in the glasse shewed by the scale) 19.5. In the truer glasse *kq* 4500 *ql* 19.6. I find the Copy of the catalogue of the Moon's semids: that I kept writ wrong in the observation on the 28 November 7^h morning: when *kq* was 4494. *ql* 21.9: whence semid: Moon 16′ 45″ which is by Lansberg 18′ 36″. That of October 19: 8^h night was observed 15′ 2″ which - 15″ = 14′ 47″ (which by the great glasse after the manner of the 3 Diagram I observed 15′ 1″½) according to Lansberge it would have beene 14′ 34″ + 4′ 2″ = 18′ 36″ but the observed 14′ 47″ + 1′ 58″ = 16′ 45″ soe that the halfe of the[13] difference 2′ 1″ - 3″ = 1′ 58″ an invincible argument that his eccentry hath[14] is rather too great. Againe observed October 9 8^h PM radius of the Moon 16′ 36″ - 15″= 16′ 21″ which by Lansberg ought to be 17′ 27″. July the 25: 8^h PM by the glasse 15′ 17″ - 15″ = 15′ 2″ when Lansberge is 14′ 46″ which + 2′ 41″ = 17′ 27″. Againe the observed 15′ 2″ + 1′ 19″ = 16′ 21″ here his 2′ 41″ againe halfed doe to my admiration differ one[ly] 1″½ which his againe is too much.

The 16 of November compared to his, & his & myne of the 28th makes his ½ eccentricity to[o] great by 10″.

These few observations that I have made since those I sent you, you will find elsewhere. I am not yet soe well furnished with observations of the Sun as that I will pronounce any thing resolutely untill I have its least diamiter, I had 3 observations of the last yeare which I cannot; If God preserve us save untill the next I hope I shall decide it. I dare not venter for feare of burning my eye to use the glasse with the scale except some seldome times when I interpose a greene glasse or 2 which hinders the light soe much I cannot well discerne the pointer to set them true & I have not yet fited a plate for it as above because my glasse is not to my desire. Such observations as I take to be neare the trueth are these. December 18 1640 suppose the glasse *k* of the 4th Diagram, at *a* of the 2 Diagram. All the lines as neare as I could measure them are thus [Table 18.2].

df.ef::90.tan, *edf* 3°.0457 therfo: *def* 86°.9543 then as
df.ef::*no.oe* 217.6 + *ni* 32.4 = *ge* 250
ic=*ik* 217.*cni* = *def* 86°.9543::*in*=*gf* 32.4 . *icn* 8°.3104
90.*ci*::90 - *cim* 5°.2646. *cm* 19.91
90 - *cin.cm*::180 – 90 - *cin.mi* 216.1 + *km* 0.9 = *ki* 217
15°.40 : 9°.95 the proportion of the refraction of this concave glasse :: *icn*

Table 18.2

The observation of the Sun's diam: with every line used in the calculation by the usuall perspect: glase	*kl* 5.5 *ki* 217 *kg* 4307 *kh* 567 *hg* 3740 *ge* 250 *hd* 51 *fe* 199 *ig* 4090 *al* 4097½

[13] Flamsteed's version has 'his' instead of 'the'

[14] Flamsteed's version has 'halved' instead of 'hath'

Table 18.3

October 31	*ge*	600
	hd	200
	kh	3430
	hg	7591
	al	4091
	kg	11021
	ig	10804
	fe	400

qcb 5.3962. *qcb* – *qcr* (=*cim*) 5°.2647 = *cbv* 0.1315
90.tan *cbv*::*bv* (=*lk* + *mk*) 6.4 . *cv* 0.015 + *vm* (=*bl*) 19.90
19.90= *cm* (=*vl*)1991
al 4097½ . *bl* 19.90:: 90 tan *bal* 0.2783= 16′42″ semid: of the Sun

The same day by a large glasse after the 3 Diag: *cd* 16048. *ad* 76.6 semid: of the Sun 16′ 25″. By the same glasse & another scale Decemb 24: *cd* 33552. *ad* 173 semid: of the Sun 16′ 25″ which without all doubt is the least & if it erre it is that way for the same day I observed it after the manner of the 5 Diag: *qk* 4495: *ql* 22.02. 16′ 52″ the Sun's rad. & according to the 4 Diag: 17′. If the weather had not forced me to use the darke box I should I doubt not in larger table then I can use in it, or any where but where the great glasse laid when you was here have found more harmonious observations I cannot use my large instrument with the crosse table after 10h morning. The observation that was in the note for the 4 Diag: & October 31: you may try over after the same sort as that above, hereby you shall find Semidiamiter of the Sun 16′ 11″=0.2697 [Table 18.3].

I thinke this is very neare the truth if it be not rather to litle, it is 1′ 36″ lesse then Lansbergs 17′ 46″ if we take this out of his greatest Decemb 18th 17′ 59″ the differ: is 12″. Also that above Decemb 18 after Diag: 3 16′25″ - that in the note October 25 Diag: 3 16′11″ the difference is 14″. This way I take to be lesse erroneous then that of the 4 Diag: unlesse we keepe the glasses alwayes unremoved & know every necessary exactly to the 100 part of an inch. Onely there are these inconveniences in this 1st it is hard to give the true diamiter of the Sun or Moon except there representation be at great distance from the glasse which though it may be performed (as thus, I find there are in all convex glasses 2 severall motions, the 1st to a point as *ABC* in Diag:1; which carrieth all colors the other of 2^{d} moves from a point as *ci* in Diag:3; suppose a glasse the ½ breadth whereof *Bi* (1:di) is 3 inch: & that this terminate the edge or limb of the Sun by the outmost BC at the length of 160^{in} *iC* proportionable unto which distance by the rules of the 3^{di} the representative diamiter will appeare; suppose in this glass the best plain wherein this apperation is made be not descidable 159^{in} 161^{in} if we contract *iB* to 1 inch it will not be distinguished betweene 157.163, yet the diamiters will be proportionable to their distances, although the light or color will not be soe easily senced to be proportionably lessened answerable to the diminution of the admittance, yet sure it is soe, & is hence easily possible soe to diminish *Bi* as *iC* may be 320 inch or more & the opticall sun apparent <~~y~~> for <~~best~~> use this can onely be practised in the Sun though the Moon first taught me it, & that the best apperition of its spots according to this reason & all experience are at a shorter *iC* then the best apperition of the diamiter, which hath more troubled my thoughts to bring into an order then all the rest. Ther is another experiment hath some coherence with this. I find by 2 glasses (Di:4) that if there be 3 Tables as *in* (2 Di) 1^{in} *hd*. 2^{in} *gl* 6^{in} that the distance of the 2 glasses *al* must be lesse to make the best representation of the Sun's limbe at *ge*, then at *hd* & not soe great here as for *in*: this may happily be some reason of the difference: Scheiner & Malapart have noted; who laboured to keepe the same diamiter of their sumer & winter observations, whose largest peripherie halfed would not include this parenthesis) by glasses of usuall perspectives, yet it is troublesome in respect it requires if we doe thus of necessity some thing to keepe all other lights from it & either a long case or ruler with a table. If you apply the 4 or 5 Diag: for the measureing of the lunar diam: the outer glasse had need to be 3^{in} wide to make the representation light enough if it be 5^{in} diami: this alsoe will be better discerned if you draw a broad black circle about it, by which you may truely find the Moon's diameter, especially neare full moon. If you use the 5 Diag: this way every apparition is the same as in the heavens, & for ought I yet see the pictures appeare more perfect then by the 4 Diag:.

I am sure it is much easier to make true convexes then concaves, & that the[s]e are in all astronomicall observations preferable before the other. I remember in my last letter I writ I thought I could shew some probability that Lansbergs borrowed his solar diamiter from the perspective glasse although I thinke it not to be true yet because those observations here inserted makes for the Sun's diameter neare his going the whole excentricyty I take it not much amisse to set them here; your Sirturus[15] shewes that all perspective grinders have forming tooles of the same axes, suppose Lansbergs or his observer, Hortensius had truely set downe. I have tryed over the latter, & find the Angl at the Sun a very few minites: by chance looking into the calculation of the latter of the Moon's appulses that I sent you I find that I missed a figure in finding * right ascension, & therefore have noted the time false & soe the rest that I sent: As soone as I have dispatched my intentions for the solar theory I shall review them & send you some other observations, because your Ephemerides is soe ready if you will set me downe some of the Moon's passages as you did heretofore especially those that will be at evening before 10^{h} I shall be as observant of them as I possible can. I expect every day a case which I will order the best that I can for exquisite observation. If there be any other notable conjunctions I will [rely] upon you for information. I have returned your Keplers, but keepe your Sirturus for a review when my Iron is finished, & let me once for all assure you that whatsoever I doe or can master shall be at your command. If you please to get me those 2 bookes Suelly obser: Hassi:[16] Kepleri harmon: mundi[17] (because I named not this amongst those of his I sent for to London) if it have any thing in it but meerely

[15] Probably *De Telescopio* by Hieronymus Sirturus (1618)

[16] The copyist has written 'Suelly', but maybe Gascoigne intended 'Snelly', in which case this is perhaps a reference to *Observationes Hassiacae* edited by Willebrord Snellius (1618)

[17] *Harmonice Mundi* by Johannes Kepler (1619)

speculation, I shall returne what you disburse for them by him I sent my last letter to you by. I was sorrowfull to see Pholicides[18] such a messenger of unwelcome tidings & shall be noe litle troubled to find an oportunity to returne an equivalent favour, it is a triple ingagement in respect of your selfe our friend & the ability of its authour, beleeve it I allmost ashamed to returne onely most due thankes; as soone as may be I shall view him over & over, he intimates what I am most certaine of am [an?] impossibility of the Keplerine hypothesis, also that the Moon descends higher from the earth then Kepler would have it & indeed 16′ 45″. 15′ 2″ gives the whole excentricity after the old manner 10800 + of 100000 but I shall discent wholely for ought I yet see from any yet extant neither allowing irrationall equall motions, nor unlikely prosthapherisis, nor the whole, nor bisected eccentricity. It is easy to order the Keplerine in a farr more facile maner; I thanke you very kindly for your labour about the Moon's difference of parallax, I see also it will be impossible by the light whereof in my last letter because the earth would produce an inequality, besides the smalenesse of the alteration which you prove for it holds in both. Time nor paper will admit a further proggresse to you most obliged friend.

Notes:

By J.Flamsteed: This letter answering Mr Crabtrees of Dec 28 is dated Jan 25 1641/40: in the entrance of it from the underwritten observations hee determines the proportion of the angle of a Ray of incidence to that angle in Water as, s, 30° to, s, 42° 15′ delivering the observations, & the angles formed by proportioning as in this little table [see letter above].

After this hee explaines severall things of which Mr Crabtree had desired a plainer account, & describes the manner of placeing his scale in the tube for measureing angles, which because wee are now furnished both with better demonstrations & methods I pretermitt haveing designed onely to excerpe his observations & some such remarkeable places as cannot with justice to the memory of the trice ingenuous author not [?] be pretermitted:

On December 18. 1640: by a telescope of a convex object & concave eye glasse. hee reaceaved the suns species on a scene & from the observation concludes the suns semidiameter 16′-42″ but the method hee used is much <defective> & because the concavitie of the eye glasse could not be measured exactly (as himselfe confesses)...

...hee describes his inclinator not at all in my opinion convenient, tho ingenuously enough contrived. & proceedeing hee addes.

> for the finding of the moones pa[ra]llax by the moon's body every time that I behold it through the diameter measureing glasse it seems more feasible then other.

his method is demonstrable but as Mr Crabtree proves not practicable: by reason that an error of 2 or 3″ in observation is inevitable which will destroy the fruit of the operation & detort it beyond the granted truth...

8.
GASCOIGNE to WILLIAM OUGHTRED
c.February 1641

Sources:

Cambridge University Library, Macclesfield MSS, William Gascoigne to William Oughtred, c. February 1641, Add.9597/13/5/104-9

Rigaud, S.J., Correspondence of scientific men in the seventeenth century (1841), 35–59 [source used]

[18] Johannes Phocylides Holwarda

Most honoured sir,

Your letter hath emboldened me to present this to your view; if the contents deserve your approbation, I shall not after it fear the most prying eye that may examine it. Your belief that there is in all inventions aliquid divinum, an infusion beyond human cogitations, I am confident will appear notably strengthened, if you please to afford this truth belief, that I entered upon these studies accidentally after I betook myself to the country, having never had so much aid as to be taught addition, nor the discourse of an artist (having left both Oxford and London before I knew what any proposition in geometry meant) to inform me what were the best authors; nor being rich enough in language to understand perfectly any tongue except my mother's, having had two years' interim between the school and university, where I dare not say that I learned more than how those lived, that increased their knowledge, and to know my own wants by others' wealth, and to be sensible that the judgment at Babel was no trifle. These difficulties might have produced despair, had I not been furnished with an ardency to vanquish them, (at least in some small measure,) and continually refreshed with a little increase, whereof the whole stature is so dwarfish, that I must intreat all your expectation of great matters (from one that is thus every way unlikely, and whose sole aim and end was his own pleasure and satisfaction) to be set apart; which hoping you will perform, I proceed.

A straight course suits not well with refraction, the paths whereof are such by-ways, that the very industrious Kepler, p.113. Paralip. in Vitellio. professeth, in genuina hujus rei causa directe et a priori demonstranda haereo. And thus marginal notes it, Dic quibus in terris, et eris, &c. In p. seq. from Vitellio's gross observation he labours, by a tedious and irrational multiplication of calculations to introduce true elements thereof. Had he in this used his usual observancy, he would have followed his old masters to whom he thus in pag. 95 testifieth his affection: plurimum namque amo analogias, fidelissimos meos magistros, omnium naturae arcanorum conscios: and would (with an admiration of this harmony) have said, what upon another occasion he testifieth in p. 650. Epit. Astron. Omnis motus naturalis vel artificialis, in quem vel eadem vel analoga concurrunt principia, dispensatur per sinus angulorum: praecipue vero et evidentissime, motus vel nisus brachiorum in libra et statera. If in the first scheme (in one of these enclosed papers [see Figs. 18.6, 18.7 and 18.8) *ne* be the handle (of a pair of balance) *e* the centre whereon the balance *gea* moves, *g* one weight, *a* another, differing (according to him) as *np*, *nf*; which are shewed in the handle and its continuation, by *ap*, *gf*. if the weight at *g* be supposed to be let fall into water, according to its proportion to the equal solid of water will be the ascension of the end of the balance at *g*, which if to *k*, *a* will descend to *c*; so that $c : a :: nd : np$, $k : g :: nm : nf$. and the angle of mutation gek $(=aec) = mek$ $(=ced) - feg$ $(=pea)$ is here not unlike the angle of refraction. For if we suppose (as I made this observation here under) at *g* a plate of well polished brass, *gf*, *fe*, a rectangle instrument of a competent size, *gea* a ray in the free air, after this *qel* a surface of water under which *ef* is placed perpendicular, *g* will ascend unto *k*, and *fg* will be (the water being to the air as when I made this) to $mk=cd$, as s.30°. to s.42°.15′. which proportion is the

Table 18.4

45°	71° 40′		s. 45°	s. 71° 58′
40	59 45		40	59 49
35	50 27		35	50 28
30	42 15		30	42 15
25	34 46	s.30° : s.42° 15′ ::	25	34 38
20	27 23		20	27 23
15	20 16		15	20 22
10	13 30		10	13 30
5	06 45		05	06 44
feg	*ced*			

same in all inclinations as is manifest by this table, in which I rather choose 30°, 42°15′ than 45°, 71°58′, because here by reason of the greatness of the inclination, and the augmentation of the obscurity of the water answerable to the secants, the brass plates could not be so perfectly discerned [Table 18.4].

If I did not believe this catalogue sufficient, I could have been more curious. For experiment sake I tried and found water 1876, glass 4791 of equal solids: and *cd* in air to *fg* in glass near 52°35′ to 31°20′, which is therefore to *hi* in water, as 31°20′ to 36°12′, whence should follow that *b* should be to *g*, as 1876 to 2915, and this in all inclinations which is not agreeable to truth, (and that *b* should be to *i* as 1 to 400 according to Galileus, mecan. p.81.) if the refraction of these could be conjoined with their weights they might have some excellent uses; therefore I have inserted this trial of their weights. Monsieur des Cartes's conceit that the rays of light move more easily in those bodies wherein they tend toward the perpendicular, than in those wherein they make from it, is not for aught I yet see without as great inconveniences as this analogy of the weight. Whether it be the one or the other, yet *cd*, *fg* are the true measurers of the refraction which he in his excellent discourse, De l'arc-en-ciel, sets as 250 to 187[19], an unit differing from mine. In all the other draughts every perfect circle is set for this, and its centre for *e*, the superficies, in which *e* is, for *ql*, the lesser angles contained by the ray of incidence (here noted by the black lines) and the perpendiculars to the refracting plane are as *feg*, and the greater as *ced*; all their sines are to be supposed where they want, and the letters in every mentioned variety to be placed in their right site, in which position *SQl*, in scheme 2, denotes a rectangle triangle of glass polished on its three edges, *CBSQb* a ruler parallel to SQ, on which ruler *CO*, *BO* are 2 perpendicular plates perforated, through which the ray of incidence *Ae* is received, and at *e* refracted to *b*, (whence all other light is to be debarred,) which marked on the ruler (or, which I generally use as the better, on a table covered with white paper and perpendicular to *SQ*) gives the measure of the three sides of *Qeb*, whence we may command *cd*, which with *fg*=(s) *SlQ* affords all the variety produced by the change of inclinations; as if

[19] *s*.42°15′ : *s*.30° : : 250 : 186.

the plates be *wa*, *va*, the ray of incidence *ate*, refracted at *e* to *h*, thence to *i*, it will be *fg* : *cd* :: *po* : *mn* :: *qr* : *ts*, (which is the same with diag.1, supposing *Ql* here, the tangent of the glass at *b* there); if *yx* be the ruler, *xi* the table, and the perpendicular to *l* fall between *Q*, *S*, the same proportion will be found in all the various inclinations. The glasses in the diagrams are distinguished by the red lines, and although I chiefly use for the angle-measuring glass, diagram *5,* yet I intend herein to begin with the 1st, comprehended by the plane *aH*, and convex superficies *akH*, wherein *AB* a parallel ray (or according to truth not sensibly differing from it) is revolved upon the half wideness *iB* at *B* refracted unto *C*, the best appearing intersection of the refracted rays and the continued axis of the glass (or the centre of the sun's idol upon a table), by some exact scale I measure this distance from *k*, and hereby have all that is required in the precept appertaining to the 1st diagram, as will appear by the other included paper[20].

Although I have not hitherto glasses according to my own desire, yet they will as sufficiently as most other London best sale glasses serve for trials, in one whereof *Wk* is (as near as I can find) 2580, *ib* 60, *iWB* 1°19′56″= *WBA, CBD* by comparing of divers observations is near 2°5′52″, whereof here is a proof agreeable to the other half of the diagram 1.

10½	*Sk*
4702	*iq*
43′52″	*iqb*
89°16′8″	*iqb* - *ibg* 1°19′56″ = *gbq* 87°56′12″=90 - *qbr* 2°3′48″
100800	*ZS*, 2′4″ *omz*, here *mt* makes no sensible difference, being near 0.0037 of 1/100 of an inch; hence it appears that *omz* 2′4″ + *qbr* 2°3′48″ = 2°5′52″ *CBD.*

Also, that *DE, GF* being any way exactly found, where the glass is of power to give *q* (differing from *C)* exquisitely, we may find *Z* which hath not in this glass, at two miles distance, 1″ difference from 0″0‴ &c., although Scheiner, p. 99, Refract. Coel., thinks there is a difference proportionable to the perig. and apog. of the Sun.

[20] Viz. "in diag. 1. having $\left\{\begin{matrix} WB \\ BI \end{matrix}\right\} \begin{matrix} BWi = WBA \\ WBi \\ Wi + ik = wk \end{matrix}$

$$wk + kC = \left\{\begin{matrix} iC \\ iB \end{matrix}\right\} iBC - iBg\left(= iWB\right) = gbi = 90 - CBD$$

Also $\left\{\begin{matrix} qi \\ ib \end{matrix}\right\} ibq - ibg = gbq = 90 - qbr$

Then *GF* : *DE* :: (s.) *mbW* : (s.) *rbq* :: (s.) *bmv* (= *tbm* = *tbW* – *mbW*) : [s.] *omZ*."

Surely the true cause of that summer and winter difference of the glasses' distance is from the rarefaction and condensation of the air, the change whereof is discerned by unchangeable glass. There is another reason, whereof hereafter.

In the first half diagram 2.[21]

37½	*Bm*
217	*iB* (as near as I can find)
6300	*kg*
637½	*Dg*

whence *HG* is to *EF* : : (s.) 15°.2449 : (s) 9°.8047, for which in the observations, hereafter recited, I use 15°.40 to 9°.95 having first found and used it: except I were more certain of all the other requisites I need not more curiosity: it may be, some seconds or decimals are something differing from the truth, because for ease I generally find the proportional part on a pasteboard double-rimmed circle of

[21] On the inclosed paper before referred to in the letter there are the following notices. NB. The arcs are repeatedly written for their sines.

$$\text{Diag. 2.}\left\{\begin{array}{l} \left.\begin{array}{r} ib \\ Bm = Il \end{array}\right\} \begin{array}{l} miB = ABE \\ mBi \\ mi + km = ki \end{array} \end{array}\right.$$

$$\begin{array}{r} kg - km = mg = \\ gD - gC(= mB) = \end{array} \left.\begin{array}{c} BC \\ CD \end{array}\right\} DBC + iBC\left(= 90 - mBi\right) = iBD$$

$$\text{Also}\left\{\begin{array}{c} lk \\ kg \\ ki \\ ge \\ hd \end{array}\right\} \begin{array}{c} cm \\ bl \\ tba \end{array} \qquad \begin{array}{r} kg - kh = hg = \\ ge - fg\left(= dh\right) = \end{array} \left.\begin{array}{c} df \\ ef \end{array}\right\} def$$

Then $df : ef :: no\ (= ig = kg - ki) : oe\ (+\ og = ni) = eg.$

$$\left\{\begin{array}{c} ni \\ cni \\ ci \end{array}\right\} \begin{array}{l} nci \\ nic = 180 - nci - cni \\ \quad = 90 - cim \end{array}$$

90: ci :: (s) $cim : cm$ :: [s] $icm : im$

$ki - mi = mk$

$[mk] + kl = ml = vb.$

$$\begin{array}{r} GH : FE :: (\mathrm{s})nci : (\mathrm{s})qcb\left(-qcr = cim\right)rbc = \\ (\mathrm{s})qcb : (\mathrm{s})nci :: (\mathrm{s})cbv : [\mathrm{s}]tba\left(= bal\right) \end{array} \left\{\begin{array}{c} cbv \\ bv = lm \end{array}\right\} cv + vm\left(= bl\right) = cm''$$

proportions. The other half of this diagram, though it shew the respect of *a* to *e*, yet because *gc* [diag. 3.] must be very short, or no distinguishing of *e,* though *a* be great light, I therefore leave it for the diagram 4, where *K* is the same with this *a*.

In the third diagram *di* is the representation of the semidiameter *af*, the true angle whereof *egf* at a very short distance becomes altogether unperceivably differing from *acf*, which is apparent if we take the observation in the paper pertaining to the fifth diagram *ql* the same of *id* here, is 19.725[22], *pf* or *fa* (1 yard) 3600, *kq* or *cd* 4533, *kf* or *ca* (11½ score) 828000, which gives *acf* 14′ 56″ and *egf* 14′ 56″, also *dci* 14′ 56″; for *DE* 2° 5′ 52″ : *GF* 1° 19′ 56″ (diag. 1) :: *dci* 14′ 56″ : *gcb* 10″ -. And *cgb* 89° 50′ + : *gcb* 10′ - : : *bc* 10½ : *gb* 0.029=*ea*. Hence is plain that the difference between *af*, *di* in a short distance may be neglected.

It is not unworthy of a note that *dc* of itself is in no length without its *di*, or that *i* is not so fixed but that *ci* may be prolonged. For if in the first observation[23] in the paper *cd* be 159 in. +, *Hb* 3, this *d* is only limited by the rays falling like *ABC* in the first diagram, which are both the colour-bringing and distance-declaring rays; if we lessen *Hb* to 1/10 of an inch, *di* will be the same at *cd* (only more obscure), and if we remove *d* to 170 inches or +, *di* will be proportionable, and *i* very distinct, which may be more prolonged (for aught I see, to the length of Tycho's or Landgrave's greatest tube) without any mixture of imperfection from the sides of the narrow passage (which mixture all observers that way have noted); only, if the glass be spotted, each spot will be the vertex of a cone, the base whereof will be in *di,* wherein their appearance is more or less according to the diminution or augmentation of *Hb*; whence the first light of this truth came to my notice, that conjoining this third and second diagram into the fourth diagram, where if *HK* have 2580 for its semidiameter *HB* be 60, which admits rays sufficient to make *ge* the sun's semidiameter, after it hath passed through the concave *lkc* of 217 semidiameter, clear and unspotted (from the glass *k*), it is easy to reserve the same glory of light or colour and yet augment *ge* (the sun's representative semidiameter received on a table), or any object received into the eye, reserving the same convex, only augmenting *HB* (provided that it should not exceed 2580; in this the hyperbola is preferable before any other figure) and proportionating a concave of a less axis, according to the necessity of the given proportion.

I believe this in remote objects to be of no contemptible use; though Monsieur des Cartes seem to say directly contrary in his De la Dioptrique, p.129, Pour ce qui est de l'epaisseur de ce verre *(HK)* elle ne peut de rien profiter, ni aussi de rien nuire, sinon en tant que le verre n'est jamais si pur et si net, qu'il n'empeche toujours le passage de quelque peu plus de rayons que ne fait l'air. All experiments duly made will confirm this possibility of the increase of objects, and therefore disallow all our narrow cases as incapable of this help. In the note of observations, according to every diagram, the sun's semidiameter observed according to this is 16′ 11″. Here is

[22] Centisimes of an inch. W.O.

[23] See these observations annexed at the end of the letter.

at large every particular, that if I have overseen myself, you may know wherein I failed. I will not say that these are true to a second, though I hope to shew before I conclude that it may be easily and certainly performed. In this there are so many requisites that it is hard to come near the truth.

Oct.31[24]	5½	*kl*
	217	*ki*
	3430	*kh*
	200	*hd*
	7591	*hg=df*
	600	*eg*
	4091	*Kl*
	Therefore 10804	*ig=no*
	400	*ef=eg - gf=hd*
	222½	*il*

ge is the sun's representation on a table, and *hd* another on a cross[25], covered with paper, and the bars drawn with circles of 1, 1½, 2 inch. rad., the lowest part of the cross is jointed, to separate it from the cursor on the ruler, when I please to view the whole sun at *g*. *df* : *ef* :: 90 : tan *edf* 3°.0136 = 90 - 86°.9864 *def*.; *df* : *on* :: *ef* : *eo* 569.3 (+*og*=*ni* 30.7) = *eg* 600; *ci* (= *ki*) : *in* : : *cni* = *def* : *nci* 8°.1213 = 180 - *cni* - *nic*; 90 : *ci* :: 90 - *cin* : *cm* 19.28 :: *mci* (=*cin*) : *mi* 216.1 (+*ml* = *bv* 6.4)=*il* 222½.

15.40 : 9.95 :: *nci* : *qcb* 5.2740 - *bcr* (= *cbv*) 0.1663 = *cim* 5.1077.

90 : tan. *cbv* : *bv* : *cv* 0.0185 + *bl* 19.26 = *cm* ; *Kl* : *bl* :: 90 : tan. 0.2697 = 16′ 11″, *lKb* semid. of the Sun.

Dec.18	*kh*	567
	hd	51
	kg	4307
	ge	250
	Kl	4097
		16′ 30″ is the Sun's semidiameter
The same day according to diag. 3.		
	cd,	16048
	2 *di*,	154- & 153+
		16′ 26″ is the Sun's semidiameter, or near it
Dec.24. diag3.	*cd*	36239
	di	173
		16′ 25″ is the Sun's semidiameter

[24] Oct.31, 1640. See Flamsteed's *Historiae Coelestis*, vol.i, p.3, where the result of this observation is printed.

[25] Two lines are here drawn at right angles, with a dotted circle, of which the centre is at their point of intersection. This probably is intended to indicate the manner of making the observation.

Jan.20.	*kh*	2460
	hd	150
	kg	10946
	ge	600
	kl	4085
		16′ 24″ is the Sun's semidiameter

It seems by these that the sun's greatest diameter is very near 33′, and, by this little, that it goes the whole eccentricity I will not be confident, until I have a greater certainty by the least diameter ; although I have observations which persuade me that it keeps very near ½ eccentricity of Lansberg.

The least semid. 15′ 53″ + 32″ = 16′ 25″ the greatest.

Lansberg's least 16′ 47″ + 1′ 12″ = 17′ 59″ his greatest.

If I had not too long trusted Lansberg and Hortensius, I should have ere this been furnished with observations sufficient to have decided that controversy between Kepler and Hortensius. At this present I am not confident of any more than this, that the greatest is about 16′ 25″.

This of the cross which I have only of late used, (and which I cannot, for the largeness of the rest of the instrument, use either after 10^h at this time of the year, or at all in the same room in summer, to have *eg* 6.00 inches,) hath certainly shewed me, that there is required a greater distance between the two glasses, to make the best representation, at *hd* than at *ge*: although I yet perceive scarce a glimpse of the true reason, yet I see this will plainly follow, that there is here some cause of Scheiner's error (before mentioned), who, as it appears, by and with Malapert his acquaintance, endeavoured to keep the same diameter, which therefore must needs have in the solar apog. a longer *hd*, or *ge*, than in solar perig., and therefore a shorter *Kl*. It may be, the reason of this hath its original thus: In my great glass, used after diag. 3, I always find that the best resemblance of the moon's fixed or the sun's moveable spots is at least 3.00 inches before (or nearer the glass than) the plane, wherein the limb is most perfect. If this in its horizontally parallel diameters had not been unchangeable, I should have doubted its fountain had sprung from some un-noted power in the atmosphere. Sure I am, this hath more pestered my thoughts than any thing I have hitherto met with; and at this present I rest thus conceited, that the first ariseth from the intersection of the greatest number or quantity of the rays moving, as in diag. 1.; and the latter from the section of the last of these with the diameter-measuring rays, which hath a kind of proof by the former related effect of the contraction of *Hb*, the perceivableness of this ceasing by the contraction.

After the old manner, without a glass, we shall find that there is more show of colour in an obscured room with a larger hole, and more perfection of the diameter's apparition in a less. If it be hard to shew some probability of the reason of this in diag. 4, we shall find it readily in diag. 5, wherein is comprehended an incredible rarity, *q* here being the same of *C* in the 1st, *ql* of *di* in the 2nd; the glass *ab* is rightly placed when *lbm* can move rightly here as *Zmbq* in the 1st: and, as there, the more remote *Z* is from *k*, the nearer is *q* to this, so here if we place an object between *q*, *k*, the nearer its approach is to *k* the nearer according to the same reason is its picture in *eg* to *x*, and contrariwise of the contrary. This is that admirable secret, which, as all other things, appeared when it pleased the All Disposer, at whose direction a spider's line drawn in an opened case could first give me by its perfect apparition, when I was with two convexes trying experiments about the sun, the unexpected knowledge. Presently, placing my eye some little nearer *x* than *m*, I perceived that the apparition was not so punctual in the eye as on the table, although I had diminished *Zk* answerable to the augmentation by the eye (for a convex added to the convex *ab* makes *Zq* shorter, and therefore *Zk* shorter,) and in diag. 4 a less convex (I mean of greater radius) being adjoined to *kc*, according to its power to diminish the operation of the concave doth also abridge *Kl*, (by which abridgment in both glasses it is very possible to prove what convex glass is of equal power with the eye), and therefore resolved that if I tried *ab* like a spectacle, and placed a thread where that glass would best discern it, and then joining both glasses, and fitting their distance for any object. I should see this at any part that I did direct it to. And so I found it (if you please to try it, the convex of a perspective and a good convex spectacle glass will let you fully see it), which the next night's trial confirmed strangely accurate and ready for finding the altitude of any small star.

Upon trial I found by the moon that *qk* (the same distance of *kC* diag. 1) remained the same, though I tried sundry convexes like *ab*, and that the moon's picture here was in every respect answerable, for apparition of its diameter, to an object of equal rad. placed here. Thence I begun to contrive for some excellent scale, which might afford me in as small parts, as were possible, *ql*, believing that art could not afford such distinct divisions as might be severally distinguishable in *ql*. At last I contrived this many barred scale, one half whereof is here enclosed[26], whereof at first view you will readily know both [the] composition and application for use. The single bar at *g* is to be let into the case, where it is to determine *ql*, which if it be supposed to be the moon's rad. augmented by *ab* to the seeming length of a yard, or of ten yards, *pf* remaining the same, we shall find the same parts in the scale, what *ab* soever we use, hence by the reason of diag. 3. *kq*, *ql*, give $qkl = pkf$. I challenge no more in this scale save the decuplation, or, if need be, centuplation of the bars, and, therefore, thus ordered of their power. I hold a quadrant graduated according to its right reason, and fitted for a mult-rimmed cursor, rightly applied to a hollow ruler, fitted with glasses and a hair, to be preferable before those that Morinus mentions (Scient. Long. p. 52, 3, 4, 5, 6), or any other at this present known to the world. I used in the sun's last eclipse one glass thus, and a little table to receive the sun's image, all on an iron ruler, moveable on the centre of a sextant of four feet, and found it very surely affording the sun's centre. After this I found by trial (what I desired in every part) as much ease as curiosity, and have here sent a catalogue of the moon's semidiameters observed by an imperfect scale, and sometimes not with a care of exactness. Such as they are, they prove the necessity of the bisection of the moon's eccentricity, as you will plainly perceive. Some other there are, which I only set here to

[26] This drawing has not been met with; but a description of the scale will be found in the paper at the end of the letter (Rigaud note)

Fig. 18.3

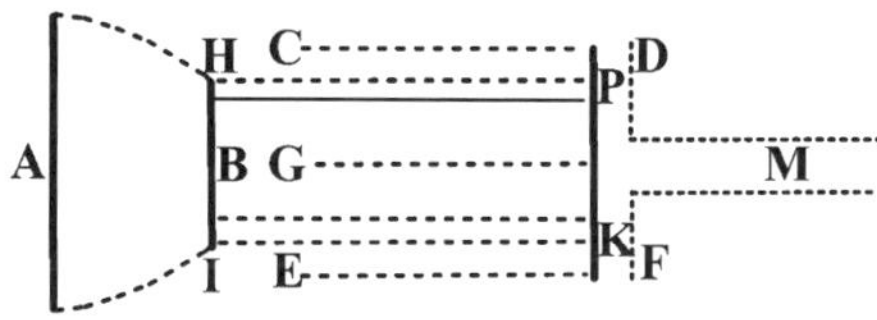

shew what may be done. Indeed I defer the making of exact scales until I have got glasses wholly to my mind, having for this purpose caused a contrivance for the certain forming of any requisite, either hyperbolic, or (according to Des Cartes) oval, convex superficies to be finished with some requisite tools or instruments for the drawing of the lines. I believe upon trial, Des Cartes's conceit will be found unuseful; surely, if I had not stayed only for an iron flawless plate, I had been furnished ere this with glasses of great command; but this country is very scarce of any skilful workmen in iron or brass, yet my desire not to be in any necessary defective hath caused me to try divers, and I think I shall shortly be fitted.

This glass in all astronomical uses is preferable before that of diag. 4, as 1st it may be contrived to admit into the eye at once 1, 2, 3° or more; the case and the glass *ab* being accordingly made in respect of the distance from *k*, whereas that only admits according to the proportion of the semidiameter of the perforation of the eye to *mc*, whence comes that mutation observed by Hortensius in viewing the moon at divers lights. Secondly, it may certainly measure this angle, so as Galileus's prodigious project in Syst. Mund. p.381 and 2, will not be necessary. In the rude draught in the margin [Fig. 18.3], suppose AB to be a plate of brass of 20′ breadth in relation to *qk* fastened to the two adjoining bars to the middle one G, and by a slit so let into the case as AB may be 10′ on either side *q*, and CGE concur with *nl* in the fixed part of the scale (whereof see the card). Then supposing the other half like this so placed as A of this concur with *b* of the other, and that CD, EF[27] may remove (as in the other card) by the help of the shank at M, A being near 40′, it will afford, when we please, without damage to our eyes, either the sun's diameter, according to the power of the scale and *qk*; or the moon's, which otherwise near the full in a wide *k* will offend the sight. AB being justly placed in *ql*, the two halves by the screws may be separated until both *B*, *b* concur with the outmost limits of the sun's diameter of longitude.

This is the easiest and most certain way that I can devise, and I tried a round plate of 154 in my greatest glass which did cover all the sun to the eye placed near M. I have attempted some trials by interposing green glass, which I find too troublesome, and therefore resolve, when I cause my cases to be perfected, to begin my observations, whereon I intend to rest, in that specified manner. If a hair be placed at right angles to the scale and pass through *q*, and be always moved in a vertical plane, in which setting it to one star, and the one of the brass pointers to another, and finding by both the pointers the distance between these stars, we have two sides, one of

[27] The lines ought to have been continued on the figure from C and E to D and F (Rigaud note)

Fig. 18.4

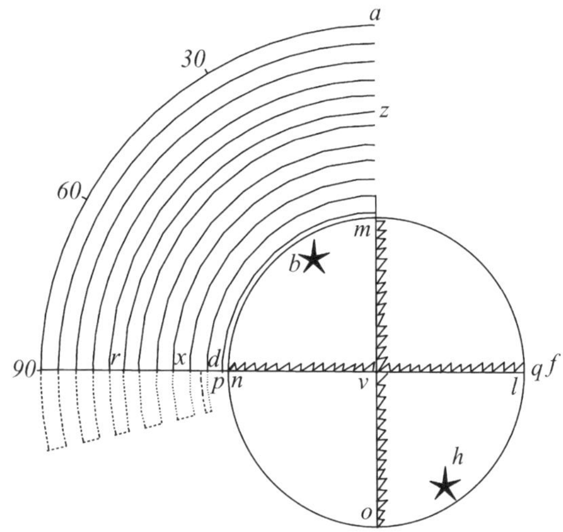

which is opposite to the right angle, and so a ready way to find the inclination, if this will content us. If not, here is an instrument may afford more exactness, though it be troublesome to fit into the case. Let *a*, 90, *dc*, be a quadrant [Fig. 18.4] graduated by transversal lines and many rims after the fashion of the fixed part of the scale, observing the true sites of each *zr* according to the allowance *ac,* 90, *d,* afford. Then fitting *rg, xy*, a moveable multrimmed brass plate, rightly graduated, and so let into *rx*, &c. as it make one plane with *a*, 90, *cd*, which is supposed on the back of the fixed part of the straight-lined scale, this cursor being fastened, by a brass pin *dp*, unto a concave cylinder *nmlo*, which is to be continued until it touch the brass pointers at *g* in the plane scale, at which end of the cylinder is to be placed a plate like *vnmlo*. One quadrant of the case *dptc* is to be so cut as the pin *dp* may move from *d* to *c* and carry with it *r* to *z*. Another slit, parallel to the axis of the case, must be made to let this pin into its true place, or the case must be there purposely jointed with a screw. If it be thus jointed it may be fitted to serve for the plates above, which otherwise must have their own length, wherein they are fastened, to themselves, which needs not to be above at most 1 f. long. When this is once placed, and *av* in a vertical position, it is ready for use. If *v*, *b*, be two stars, by any little pin in the cursor I move *rx* and its parallel *nv* until *nv* concur with the star *b*, the other being in *v*, therefore I have *evh* 30° the true inclination, because this glass shews all by contraries to the eye and right on a table. I had almost forgot that the graduations of the quadrant are to be continued from 90 to the line *yg*. If the outmost length of the case be made of wood, well squared toward the midst, and fitted to a standard perpendicular to the horizon, in which the upper part may be turned about, it will at all times be ready for use. It is also requisite to graduate that length of the case, that is next to the eye with a spiral line, divided by equidistants into what parts will best agree with the diameter-measuring scale for the ready finding of *qk*. It will be better if the case have two lengths than more.

The use of this is not comprehended within one degree, two, or three, but may by two observers and two dioptraes thus fitted with glasses, hairs, and moveable rims, be made serviceable for all angles, and give more than every second or decimal answerable unto them, which is much readier. Such scales were needless heretofore, when, by the confession of the best artists, there lurked at least 15″ or 30″ error even in the sun's best observations. I believe also by a ruler with a hair in it, moving upon the centre of a circular instrument graduated with transversal lines and two glasses, *ab*, *k* joined at a fit distance to the moveable ruler, and *k* placed in a perpendicular to the long ruler *km*, in a room not unfit for the purpose, we might in a few nights find the true meridian and pole, and, fixing the instrument there, have the star as a ready means any night to try the difference of ascensions between this and another, either planet or fixed. If the night be too dark to let us see the hair, a glimpse from a candle will help that, and not take away the use of the instrument. This I found in measuring the distances of stars in dark nights by the pointers.

If we allow that it is possible to have the moon in its greatest variation of altitudes, in its two limits in the meridian, to bear in any point of its visible hemisphere the same respect to the centre of the earth, it will follow that, from its own body, by this glass we may find its true parallax of altitude. August 1640. 25. I observed, diag. 5, Jupiter 51″ diam. Mars 38″. The glass not being potent enough to separate the two attendants of Saturn from its body, I could not measure it. Dec. 24. Mars 25″, Venus 25″-, as near as I could discern their limbs by such ill-suited glasses as I yet am worth. Jan. 11th[28] the moon's lower limb was 30′ 38″ (the moon's diameter) above the northern eye of Taurus, the moon being very near the meridian, and its centre near 2′ southward of the star. Feb. 9, 1641, the moon's northeastern limb and a star, praecedens duarum in colobio Orionis, (whereof I yet had no time to calculate the long, and lat.) were 15′ 2″ distant, the star vertical being some 4′ within the moon; they were 16′ 42″ separated at 36° 40′ alt. Cord. Leonis, the moon's eastern limb and the fixed above it in the same vertical; alt. Basilis. 39° 24′ dist.19′ 2″; alt. Cord. 44°. 40′ dist. 30′ 54″ (=diameter of the moon when by a dial the moon shewed 1½h. —). These altitudes were taken by a quadrant of no more than 1 foot radius; the elevation of the pole 58° 41′; our long, by the maps near that of Oxford.

Surely this glass will produce in a short time many exact observations for the finding of longitudes, if the moon's theory were within a few minutes true. It will be of no small use for the correction of that and the observation of its eclipses, &c.; as also for the leaving perpetual testimonies of the other planets' situations to any known (or if we please as yet unnoted) fixed. But I know I need not, when once you have a glass thus fixed, to tell you a word of its use; and therefore, only adding two other as yet unnoted, I will conclude. The first is, that by two such as *ab* you may augment any small object placed near the one of them by the same reason of this, and if you please by the scale give the proportion of any part. If they be placed in a long case, and the object *Z*, diagr. 1, be so ordered as it may be light enough, it is an

[28] This and the following observations are inserted by Flamsteed in his Historia Coelestis, vol.i. p.3. In the first he reads 'occidentalior' in the place of 'southward.'

admirable representer as well as a wonderful augmenter of the species, which appear much more perfectly coloured than that of Des Cartes, p.132, de la Dioptr.; only, as all other idols, they are pictured on the table covered with paper in their right site, and on the bottom of the eye representing or represented by that, which position, a man being in an erect station, is changed in respect of the horizontally equidistant station by the digression of the optic nerves after their conjunction, and by their reflection in their approach toward an embracement or mutual respect presenting themselves with their carriage into almost the midst of the great cells of the brain, where the ends have a contrary abode in regard of the vertical posture, as will appear to any that trieth it. The second and last is, that a third glass like *k* presenting an image in the place, where the last-mentioned small object was to be, it will appear to the eye right, and affords two planes for the scale, yet every apparition in greater obscurity than the other, which may be easily proved by the precedents.

If you judge these worthy of a public view, I shall labour to reduce them into the best order that I can, and to add some little more to them. I have shewed some part of these to two of my very late acquaintance, both industrious astronomers, who are not a little taken with them, and the more, because they believed they had tried all possible means, that glasses could afford for the measuring of diameters, yet never attained to any more than guessing, and indeed were so confident, as they would not believe until I let them see a glass to try the moon's diameter by, and how I proceeded in the other usual one. Indeed if one of my noble friends had not, before I intended it, bolted out so much as it could not longer lie hid, I had reserved it until I had got some rare glasses worthy of a prince's inspection, unto whose notice, at his majesty's being at York, he had, unknown to me, promoted this invention, who, as I conceive his words, will expect an ample relation. He also acquainted me with the famous Sir Kenelm Digby, unto whom I first shewed the contrivance of the scale, and its use in a glass, whose courtesy bound me to fulfil his request for some of the diagrams and observations, which I have lately sent him.

Good sir, do me the favour to let me know truly what you find upon trial, comparing these with what is already known, to which end I have largely and truly made relations of all these inclosed observations. What concerns engineering or aught else, every one according to their own employment will be ready to practise. That I lose no credit is the sole gain I expect; and that the lovers of art may know the advantage that this will afford, is the only end wherefore I would divulge it. If I add only a hair breadth to the knowledge of others, so it be useful, it will content me. If you please to let me know your mind, your letter at London may be delivered to Mr. Henry Gascoigne, at the sign of the Three White Lions, in the upper end of Paternoster Row, who will take care that it come safe to,

Worthy sir,
Your most obliged friend,
WILLIAM GASCOIGNE.
Middleton, between
Wakefield and Leeds.

OBSERVATIONS SENT WITH THE PRECEDING LETTER.

In diag. 3. $\left.\begin{matrix} cd \\ id \end{matrix}\right\} icd$

$DE : GF$ (in diag. 1.) :: (s) fge = (s) icd : (s) hgc = (s) bcg. In this diag. $\left\{\begin{matrix} bcg \\ bc \end{matrix}\right\} bg = ea + ef = af$

By our large glass (suppose) *kcl* Oct. 25, I observed $\left\{\begin{matrix} cd & 159.39^{in} \\ di & 0.75^{in} \end{matrix}\right\}$ 16′ 11″ *dci* sun's semid.

In diag. 4. $\left\{\begin{matrix} kl & 4091 \\ bl & 19.26 \end{matrix}\right\}$ 16′ 11″ *bkl* sun's semid. Oct. 31.

I prefer this before the finding of *tba* by *cbv*, as less capable of error.

In diag. 5. $\left\{\begin{matrix} kq & 4362 \\ ql & 20.5 \end{matrix}\right\}$ 16′ 10″ *qkl*, Oct. 25 sun's semid.

This observation was made by a green glass to obscure the Sun.[29]

Also diag. 5. $\left\{\begin{matrix} kq & 4533 \\ ql & 19.7\frac{1}{4} \end{matrix}\right\}$ 14′ 57″ *qkl*.

$\left.\begin{matrix} kf & being & 11\frac{1}{2} & score \\ pf & & 1 & yard \end{matrix}\right\}$ 14′ 56″ *fkp* the true angle

This observation was made for the trial of the glass, as also for the correction of another glass whereby I had made most of the observations of the moon's diameters, whereof I found the convex next the eye was removed out of its due place by drying it, when it was in cold nights misted by the eye, which inconvenience may be avoided by laying either the whole, or that part next the eye, within the warmth of some fire before it be used. I first tried this glass in a glorious sun, wherein 2*ql*= two yards were precisely terminated by white paper, and *kf* 11½ score yards carefully lined out; at which distance in the next dark night two candles were halved, hereby proving that the dilatation of the apple, or perforation of the eye, hath not

[29] These three are the observations, of which Flamsteed has inserted the results in the Hist. Coel. vol. i. p.3

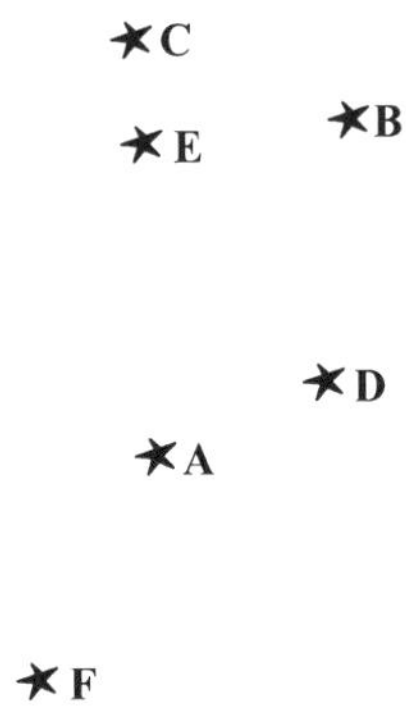

Fig. 18.5 The Pleiades

that operation in this as in diag. 4, and therefore not liable to that error, whence ariseth Propri. 5. Hortens. p. 44. Dissert. de Mercurio in Sole.

In the same book, p.52, he makes	AC 40′ AE 31 EB 18 aut 20 ED 30	which I find	AC 38′ 18″ DF fere = AE 27 50 EB 19 24 ED 25 31 AB 35 1 BD 22 43 AF 22 54 CB 21 O DC & AB fere =

I intend not here to set any thing but what I affirm is not so perfect as I resolve to rest upon. These by BD, ED, CD, will evince that for the present I am content with some error, yet I am sure ABED are very near the truth. I might also have placed many others adjoining to or contained in these, if I did not credit these sufficient [Fig. 18.5].

THE DESCRIPTION OF THE SCALE ON THE CARD, AND ITS USE.

The line *e* is 11 of *d* divided into 10. The other bars afford a decuplation of intersections performing as much as *d* and *e*, ten times their length, would do, by 101 of *d* = 100 of *e*. So as if these [marks[30]] on the moveable *ge*, *sb* be brought in one straight line with their like to the fixed *nlgr*, we shall find 100+ on *d*, 40+ on *e*; 4 on *f* for 144, the distance of *g* from the axis of the case. This moveable should be made of brass, and all the lines drawn as small as may be, and let into the fixed, that both have one common plane, if all be finished by a skilful workman. By a convex glass

[30] Representations of them are here roughly drawn in the original.

we may increase the scale, that we may discern + or - and make use of them, or if need be of more bars. A screw must be so contrived as it may remain parallel to *sb*, and being screwed through a staple near *ac* on the fixed, it may remove (by another standard near *s*, wherein it must be so fastened as it may turn about), and carry the moveable by its motion in and out. The other half must be ordered agreeably, and both *gg* meet at the axis of the case, and *d* continued.

CATALOGUS SEMID. LUNAE.

				h
1640	Aug. 25	15′	17″	8 p.merid.
	Sept. 19.	15	11	8 p.m.
	Oct. 9.	16	36	8 p.m.
	10.	16	36	8 p.m.
	14.	15	54	8 p.m.
	15.	15	34	8 p.m.
	16.	15	28	8 p.m.
	17.	15	23	8 p.m.
	19.	15	2	8 p.m.
		15	1	diag.3.
	27.	15	38	7 a.m.
	29.	15	41	7 a.m.
	30.	15	43	7 a.m.
	31.	15	49	7 a.m.
	Nov. 9.	15	49	6. p.m.
	10.	15	40	6. p.m.
	14.	15	11	6. p.m.
	* 16.	15	14	6. p.m.
	17.	15	21	6. p.m.
	20.	15	18	6. p.m.
	28.	16	45	7. a.m.
	Dec.1	16	18	7. a.m.
1641	Jan. 11.	15	19	6. p.m.
	14.	15	41	6. p.m.
		15	23	(diag.3.) 6. p.m.
	16.	16	0	6. p.m.
		15	40	(diag.3.) 6. p.m.
	Feb. 9.	15	27	6. p.m.

*This 16 day I tried the moon by the glass, whereby I made the observation at 11½ score, and found that the diameter was 31″ less than by the glass I had made all the former observations by. Therefore it seems they are to be 15″ less than are here set. I also found the distances of the Pleiades formerly observed agreeing hereto. And therefore made all these after by the tried glass.

Fig. 18.6

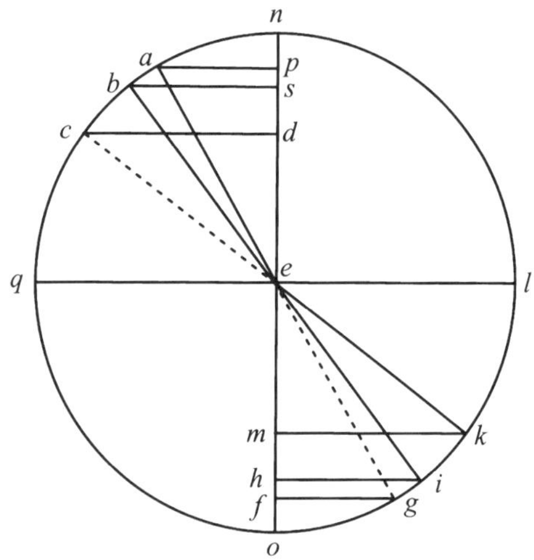

Scheme 1

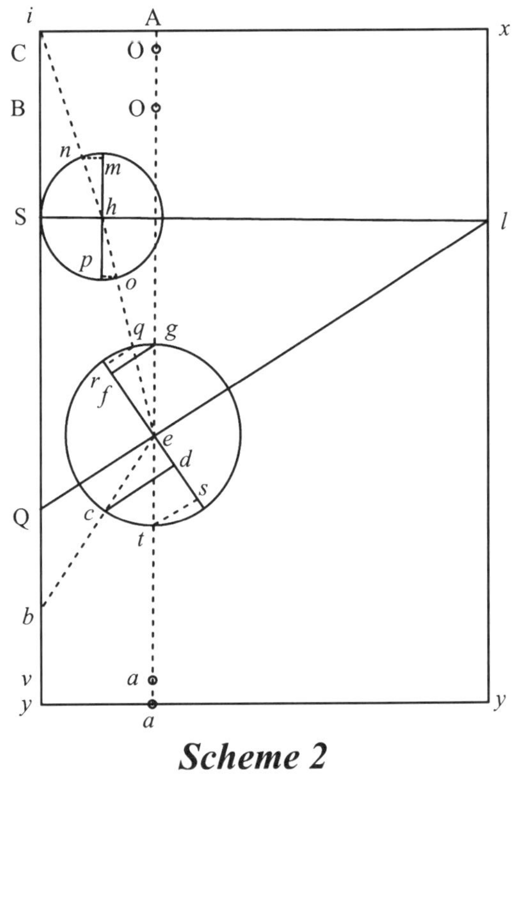

Scheme 2

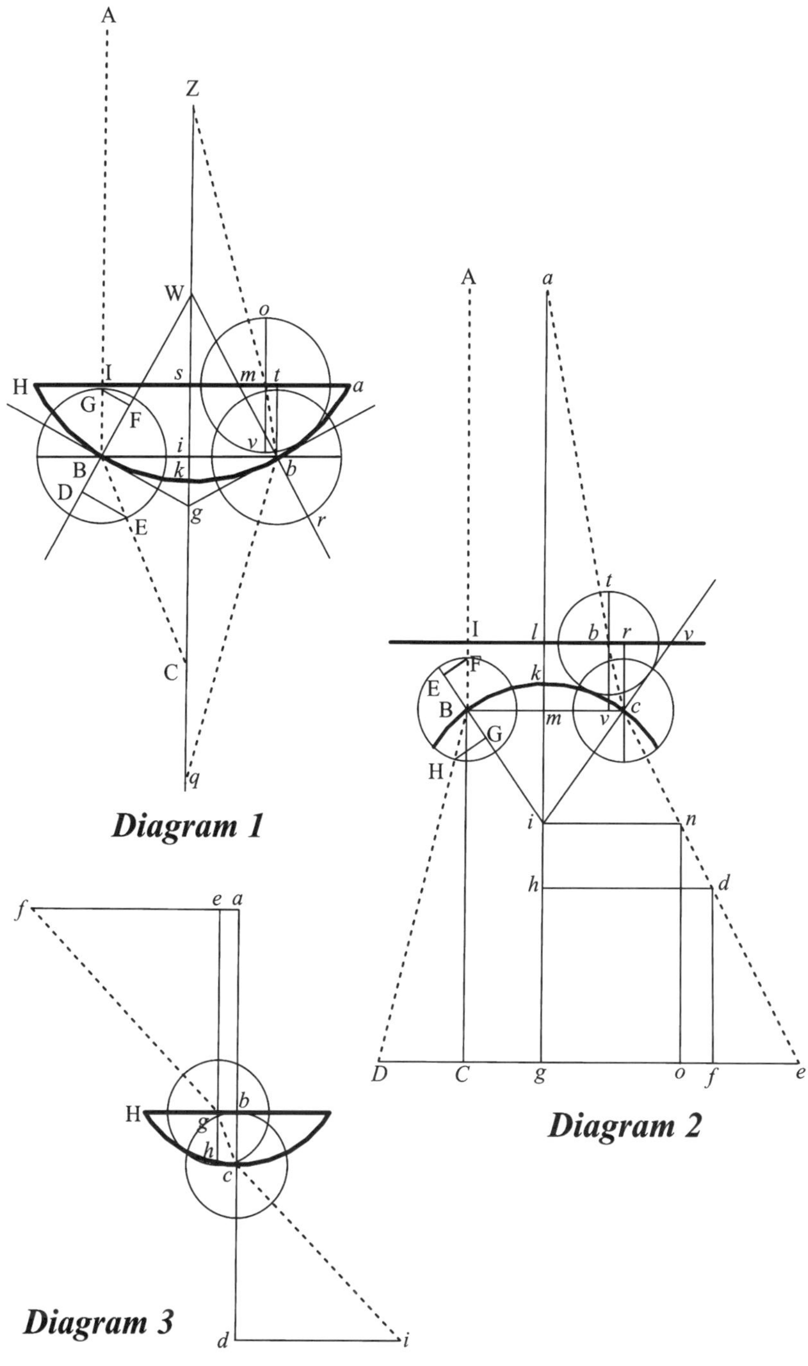
A
Z
W
o
H
I
s
m
t
a
G
F
i
v
B
k
b
D
E
g
r
C
q
Diagram 1
A
a
t
I
l
b
r
v
F
k
E
B
m
v
c
G
H
i
n
h
d
D
C
g
o
f
e
Diagram 2
f
e
a
H
b
g
h
c
d
i
Diagram 3

Fig. 18.7

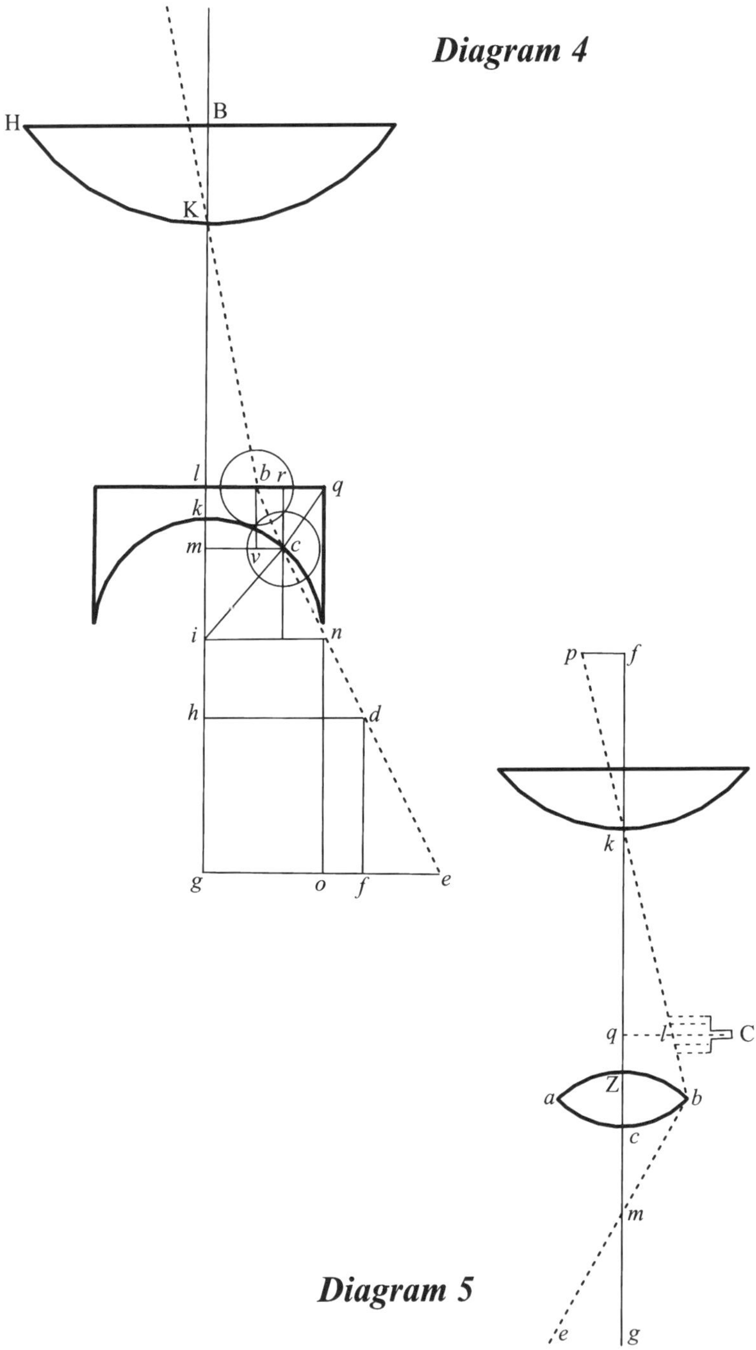

Fig. 18.8

Here are three more which the time will not permit me now to calculate[31]:

Feb.	11.	21.1¼,	Rad. Lunae ql,	4485 *qk.*
	13.	21.5		4489
	16.	21.8		4488

9.
WILLIAM CRABTREE to GASCOIGNE
18 March 1641

Sources:

The National Archives, RGO 1/40, ff.13v-14r
Bodleian Library, MS.Eng.7031, ff.29r-32r [source used]

I pray you pardon me this time. Your application of Keplers conceit concerning the weights in a paire of ballance to refractions in Aire & water is a pretty conceit if it would hold, but for my part I could never tell what to make of that place in Kepler; either I understood him not, or els he is out, I pray you let me have your second thoughts about it, For let *ne* be the handle of a pair of ballance or gold weights *ql* the beame equally poized by the weights *r* & *s*[32] Now I say let the weights, be increased never so little, & (as I conceave it will draw downe that end of the beame to, *o*, & not rest in, *g*, nor in any other intermediate position, unlesse the letter bee in the pin, *e*, or the stringes of the higher ballance staying upon the beame,

If you try it exactly I thinke you will find it soe. Also where you say the weight at *g*, being let into water, *g* will ascend to *i* according to the proportion of the solides of the weight & water; I am a litle stumbled what you meane by the proportion of the solides of the weight & water, but I conceave you meane that the said weight at *g*, being let into water *g* will ascend towards *i* more or lesse according to the proportion of the weights of (any quantity) suppose an inch of water be to an inch of that matter whereof the weight is composed; If this be not [the] meaning I must intreat a word more of explication. But whether Keplers conceit of the ballance prove true or false your proportion of the refractions in water seemes very probable, & that the refractions there are exactly determined according to the sines of Angles. I am also verie well satisfied with your wayes of finding the sines DE, GF (which regulate the refractions of all inclinations in the same glasse) by a triangle or prisma of glasse.

You have taken a great deale of paines for mee in explaning these demonstrations to drive them into my unprepared conceit, I was a meere stranger to the very first Notions of these things before I received them from you, But I thinke I should

[31] The last memorandum seems to indicate that the letter was written in Feb. 1641. A continuation of these measures of the moon's semidiameters may be seen in the Hist. Coel. vol. 1. p.5

[32] The transcribed letter has an empty space here as if the transcriber intended to copy a diagram of the balance from Crabtree's letter, but omitted to do so. This makes it difficult to interpret the use of letters for points on the diagram.

understand perfectly the meaning & ground of all the lines in the Diagrams you sent me at first, if I were cleared of some doubts about the faling of the beames; which I especially stick at the 3, 4 & 5 diagrams & must crave your direction a litle before I can come of.

In your first paper of directions you say, *fge* (in the 3 Diagram) is equall to *icd*. In your last letter you say thus (I thought I had somewhere in my former letter shewed how *icd* in the 3 Diagram was said to equall *fge* + *gcb*[)]. Now for as farr as I conceive the Angle *fge* should it selfe equall *icd*, (as you writ first) for *mn* the tangent of the point *c* being parellell to *gb* (& the refraction of each point in a Cur[v]e being regulated by its tangent) & so the beame *fgci*: being refracted as in two parrellell superficies in the points *g* & *c* must needs (as I conceave) make the Angle *icd* at its coming out of the glasse equall to *fge* at its going into the glasse.

But still I understand not how the Beame *fg* comes to be just refracted to the midle of the glasse; for suppose In the following Diagram[33] *pq*, be the widenesse of the hole of the cover of the glasse through which the beames are admited to the glasse, Mee thinkes *fp* the outmost beame that come to the glasse should be refracted to some other part of the convex superficies then just to *c* as perhaps to, *o*, the same doubt troubles mee about the 4 & 5 Diagrams, how the beams *kb* & *kp*, which measure the Angles at *k* should come just from the point *k* the middest of the convex glasse there, & not rather from some point wide of that, as I said before.

I beseech you therefore as you have begun with mee, take so much paines more as to make me understand mine error herein, as plainely as you can for those things which by reason of your versednesse in this way are common & easy to you, are by reason of my strangenesse therein hard & obscure to me. If there be any thing in your former letters which will readie me, if you but explaine it & refer mee to it, I shall spare no diligence in comparing, For though I am not like to get requisites to make use of those things, or but few of them, yet since you have accepted mee for your scholler I would faine learne my lesson perfectly. Neither shall I be ashamed to show my ignorance freely; For though you seeme to explaine it, & Tarde often affirmes it, That all the Suns rayes are perpendicular to the glasse *HS*, or *Bi*, in the 1st Diagram you sent me; yet my judgement is not convinced but that the outmost beames of the Sun shining upon *B* fall (sensibly) obliquely upon the glasses superficies *Bi*, when it standes perpendicular to the Sun's center. Because the Sun (notwithstanding his great distance) is of a sensible magnitude & subtends a sensible Angle; & sending out beames as well from his outmost edges * (*f*), as from his center * (*a*). those beames coming from his edges * (*f*) fall upon the open part of the convex glasse (*pb*) in forme of your segment of a Cone: not sensibly differing from a perfect Cone; (The apparant semidiamiter of whose base (*ef*) doth not sensibly differ from the Sun's apparant semidiamiter (*af*) because the bredth of the glasse (*pb*) (the lesser base) beares no sensible proportion to the Sun's distance (*ab*) & soe onely the beams coming from about the midle of the Sun (as *ab*, *ep*) fall perpendicular upon the convex glasse; & those proceeding from the solar limbe make an Angle

[33] The diagram seems to be missing in this transcript, though a space has been left for one.

(*epf*) with those parrellell rayes, sensibly equall to the Sun's semidiamiter, & therefore (as I conceave) those rayes of the Sun's limbe are refracted both in the plaine & Curve superficies of the convex glasse; which will alter the proportion of *DE*, *GF* (in the 1st Diagram) making *GF* greater then the way you teach mee in your last letter by finding the point C, in that 1st Diagram. Soe that if this be thus one cannot find the proportion of *DE* to *GF* by the distance of C from the convex glasse, but must search it by the prisma of glasse.

And here also I cannot satisfie my selfe in another thing concerning the falling of the ray *fgci* in 3rd Diagram (& the like in your 4 & 5) (viz) how that ray after it have passed through the convex glasse makes the species of the Sun, *id*, bigger & bigger the further it is distant from the glasse. When (as I conceave) the Cone of the Sun's species (not withstanding his refraction in the convex glasse) is still lesse & lesse until it come to the clearest point (C of the 1st Diagram) & so shines (as for example) upon the concave glasses of our ordinary prospectives) with a lesser circle then it passed through the open part of the convex, which seems to contradict the 3rd, 4 & 5 Diagrams. These are the difficu[l]ties which I stick at, & I doubt shall not get of without helpe; if I understood these I could give a reason of all the rest in your Diagrams. I conceave there w<ith> some maine principle in these refractory refractions which I am yet deeply ignorant of which is the cause of my predent [present?] doubts, I doubt not but you will easily perceive what & where it is, I pray you therefore helpe me out of some easie explanation. I should desire you would draw me a picture of the Sun's Cone at large as it passeth refracted through both the glasses of the tube after the 5 Diagram: Or but two or 3 rayes in that solar superficies (viz the parellell rayes *ab* & *ep* in the Diagram of the precendent page) & the oblique ray *fp* that comes from the side of the Sun) according to the wayes they make, & as the[y] are refracted in both the glasses of that 5 diagr; that so I might see & understand how those rayes goe & mingle & concur, I should then perfectly understand all the rest in your diagrams & demonstrations, for I conceave I should by that see it demonstrated how the Sun is represented (through a convex glasse) situ everso; & how the Sun comes to be most perfectly idolized at *C* in the 1st diag: & at *q* in the 5, & to make a lesse image there then in any other distance; & why yet there it concurs not in a meere point but hath such a bredth (as *ql*) which compared with *qk* its distance from the convex (& making that distance Radius) will as it seemes subtend the angle *kl*=the Sun's semidiameter: Those things cleared I doubt not but I shall hereafter understand you without puting you to any more trouble in explaining principles. .... One thing more I had forgotten, Tarde p 59: Prob: 20 saith. Radij per medium Diaphanum effluens, si oblique in aliud Diaphanum majoris densitatis inciderint franguntur a superficie secundi Diaphani ad lineam perpendicularem.[34]

The same I find true in water by common experience & by your first Scheme in your last letter; & the same is alsoe excellently proved in glasse by your 2nd & 3rd

[34] Tarde, Jean, *Borbonia Sidera* (Paris, 1620), 59. 'If rays travelling through a transparent body meet obliquely another more dense transparent body, they will be broken on the surface of the second transparent body, & being broken will incline to the side of the perpendicular'

Schemes of the triangle or prisma of glasse. The same course you hold with the beame *z* & your 1st Diag: & *fg* in your 3rd Diag: but in your 2nd Diag: the line *ab* (& in the 4th Diag: *kb*) being refracted in the point *b* of the plaine side of the concave glasse, you draw as if it were refracted the contrary way (viz: a linea perpendiculari) as you may see if you have the draughts by you. I suppose it was a mistake in the draught.

I pray you informe me whether you find the burning point *m* of the glasse, *ab*, (in your 5 Diag:) as you taught me to find the point *q* (wher the Sun is more perfectly idolized) by laying aside the other glasse, or by what other meanes. It seemes the superficies of the glasse *ab*, ought rather to be composed of segments of Ovalls then of Circles, Mydorgius the Frenchman hath lately written a booke in latine concerning Conicall sections which he esteems a fundamentall one to Catoptrickes & dioptrickes, but applyes it not, onely promiseth to doe it hereafter in other bookes to come forth. I have sent for the booke.

It seemes the[y] at London cannot give there glasse any certaine for but by ghesse, & therefore I have not yet sent to Mr Towneley concerning the forme or proportion of them. For ought I know I shall goe my selfe to London before Easter, & then if I did perfectly understand the ground of these things & the directions you gave me for the formes in your last letter I would make a tryall what would be had there. I could draw a patten of the glasse *a,b*, 2 inch broad, & the axis of its opposite superficies to be as 2.3 to 13.2 such as the thicknesse 0.3: as you give me the proportions (but I know not whatt bredth or convexity is best for the glasse *K*.) It seemes the Angle which one may observe by this kind of glasse is more or lesse according to the bredth of the glasse *a,b*, which if it will represent the object as cleerely is a most notable device; I pray you let me know whether you think it will or no.

It seemes also that the widenesse of the glasse *k* should be more or lesse according to the convexity of *a,b*; For such an *ab*, as will make a great angle after the conversion at *m* it seemes requires a proportionable widenesse of *k*. The reason whereof I doe not understand, I would faine know in case the glasse *k* be not wide enough, whether it onely dim the representation, or lessen it, or both, As if *k* were one inch wide, *qkl* or *pf* 15′, which the conversion at *m* were 6°, & I had another *a,b*, of power to double the former augmentation of *pf* 15′ & make it after the conversion at *m* 12°: I would know what the inconvenience is I let the same widenesse of *k* (one inch) remaine as before; whether the representation bee dimmed or lessened & how.

You say also (in the former example) that the Cube of the sine of 6° to that of the sine of 15′, so is the augmentation of the solid by *a b*, above *ql*; I suppose the proportion holds as well in lines, thus, as the sine of 6° to the sine of 15′ so is the augmentation of the line *pf* by *a,b*, above *ql*, if I be out in this I pray set me right.

You give me directions to get an *ab* of about 2 inch wide & the burning point thereof some 3 inch (I suppose you meane *m* 3 inches distant from *ab*,) and soe betweene *k* & *ab*, about 50 inches. Or if I had a wider *ab*, all other thinges answerable (the most usefull being those that have the burning point *m* least remote) but you say the glasse *k* cannot be too wide if I have an *ab* of 3 inch from *m*, & of 2, & of 1, (suppose you meane here 3 severall, *a,b*, of such severall convexities as have their burning point in those severall distances of 3, 2 & 1 inch from the glasse.) but you say to get this at London is impossible except one had a case of purpose & by trying themselves all that one could conveniently meet with, accidentally one might stumble of 2 or 3 such as would indiferently agree. Now I doe not fully understand you here in what the difficulty consists, whether you meane that it is hard to get two or 3 such *ab*, as have their burning points so remote as aforesaid: or you meane it is hard to get a glasse *K* to agree with the glasse *ab*; & how to find this out by trying I cannot well tell, but I should thinke it were best first to try the distance at *m* the burning point of *ab*, from that glasse, (if one can try it by it selfe as you taught me in the other convex) & then to try the distance of *q* from the glasse *k* & to place *ab* neare as farr from *q* as it is from *m* (for in your first directions you say *q* must be the distance from the glasse *ab*, that the eye placed nearer them *m* discernes any object most augmented & perfect). Now the proportion of the convexity of the glasse *k*, to that of the glasse *ab*, it seemes its arbitrary, for it seems by the former example that the same glasse *k* may serve to 2 severall *ab*, that will multiply the object one double to another, yet I should desire to know what widenesse & what convexity of *k*, you thinke will represent the object best or sufficiently, with such an *ab* of about 2 inch broad whereof you gave me the proportion of the convexity as befores & whether the too great widenesse of *k* will not let in too many beames or too much light into the glasse & soe confound the sight. Now I pray you where you see stick helpe me & whereof I am too short in concerning your full meaning instruct me a litle further; It vexeth mee that I cannot apprehend your meaning more certainely & punctually; but I was so meere a stranger to those notions which you were so well acquainted with that my ignorance doth sometimes make too vast a difference betweene your expressions & my understand.

But if you would be pleased to satisfie me in these doubts at this time I doubt not but I shall then rightly understand your grounds & be afterward able to understand your meaning with litle adoe.

When you have your glasses to your mind I should wish you to observe the Pleiades againe, because there are some small differences of the place of *c* found by *ac*, *dc*, *bc*, according to your former distances.

I thought your instrument for forming glasses should have beene cast of Iron or at least made of cast Iron. If forging will doe it, we have a Gunsmith some 3 miles from mee (with whom I have some acquaintance & could have him doe any thing for mee) who for complait & solid workmanship & especially in that everie point of forging is thought (by many judicious workmen) to be second to none in England; if his best endeavors may any way stead you assure your selfe I will take course you shall not be awanting in any thing you desire. About a fortnight hence I shall goe to Wigan & then I shall [see] what they can doe about casting brasse.

I thanke you for puting your demonstrations of the 4 Diagram into numbers by examples of calculation of the Suns semidiamiter: (examples are the life of instruction) I could now imitate you in the like if I had a glasse of a right forme, & doe understand the reason of every thing but the falling of the rayes & those other doubts before mentioned in which I must intreat your direction as before.

Scheiner in very many places is still upon it, that the distance of the object alters the clearnes & greatnes of its representation, if the glasses be kept at the same extent, I suppose indeed it will hold in small distances, but not in such vast distances as the Sun; but that strange experiment which you mention by the tube of the 4 Diag:, that the glasses must be nearer together to make the clearest representation of the Sun's limbe at a greater distance of the table then at a lesser distance, I suppose will fully answer Scheiners Conceit concerning the altering of the distance of the 2 glasses in respect of the Sun's approach to the earth, considering that he still tooke the Sun's image in one Circle, both in his winter & summer observations, & soe put the table nearer or further from the glasse till the Sun's image filled that circle (as appeares by the place quoted) he beates upon this in very many places of his rosa ursina, that the distance of the object alters the representation. take one for all; Lib:2. Prob:7: p:135: si objectum seruat eadem sensibilem distantiam a tubo, facile est eandem sub magnitudine eadem picturam exhibere. retentis etiam omnibus distantijs immotis habetur intentum. Si obiectum discedit sensibiliter, certum est immotis omnibus alijs, tubo, lentibus in tubo, et charta, imaginem non tantum minorem, sed etiam confusiorem pingi, quia obiectum sensibiliter abscedens, proijcit conum post conuexam breuiorem, vnde caua respectu huius imaginem minorem, et ad chartam immotam confusiorem transmittit; cui rationi suffragatur perpetua experientia. Remedium est, vt lens caua ad conuexam accedat, donec imago aequalis et formosa in chartam affluat. Si obiectum ad Telioscopium sensibiliter accedit, certum est augeri picturam ob rationes contrarias prioribus, quia concursus ultra cauam prorogatur, et sic caua conuexae vicina imaginem ampliat in chartam, adeoque non debite distantem confundit &c. cuius contrarium fiet, si caua a lente conuexa sensim discedat, donec omnia vt prius limpida et aequalia euadant. Et sic eiusdem magnitudinis picturam retinebis, in distantia quacumque &c. Et hoc artificio sub eiusdem Circuli obseruatorij aequali capacitate, excepi semper tam brumalem, tam perigaeum, quam apogaeum Solem.[35]

You commend the single glasse of the 3 Diagram above the 4 Diag: for taking the Sun's semidiamiter, but then you thinke it convenient that the representation be at a great distance from the glasse, now I understand not that the proportion of the Image to the distance will hold to find the Sun's diamiter at any place but where the Sun is most clearely represented (which as you taught mee is a litle from the end of the case of our ordinary prospectives Tarde affirmes the same, & I find it so by triall) but my want of glasses makes me that I cannot try whether such a great glasse as you suppose (the ½ bredth whereof *Bi* (1 diag:) is 3 inches) make that clearest representation soe far from the glasse as 160 inches. I first doubted whether by *C* here you understand the same thing as before in your letter, but it seemes you doe because you say the best plaine wherein the apperition is made may bee not descidable betweene 159in, 161in & that if *iB* be contracted to one inch it will not be distinguished 157in, 163in, where it seems yet that the very best representation falls (in the midle) about 160in.

In our litle glasses the beames in the point *C* (as I understand *C*) are soe vehement that one can scarce indure to looke upon them for the Sun's diameter, nor cannot (I beleeve) see his spots in that litle respendent compasse. That also which makes mee somewhat doubt whether by *C*, here you meane this clearest place of the Sun's representation, is because you say if this bee performed by glasse of ordinary

[35] Scheiner, Christoph, *Rosa Ursina sive Sol* (Bracciano, 1630), book 2, ch.33, prob.7, 135

perspectives it is troublesome in respect it requires of necessity something to keepe all other light from it, & either a long case or ruler with a table, now why this should be more difficult then the ordinary way of observing in a darke roome, I understand not, soe that there is some matter in those passages which I thinke I doe not fully apprehend. That device whereby you show that Lansbergs might probably have gathered his Sun's diameter from the perspective glasse might well have cosened as great a wit as his; & would carry a faire show of truth to any that were not exactly acquainted with the genuine nature of refractions.

Your inclination is a very ingenious device, I kindly thanke you for its description, I supppose it will require a great deale of care & exactnesse both in the makeing & useing, but if right glasses could be gotten, these other difficulties would bee but eager pastimes till we had them.

Notes:

By J.Flamsteed: Mr Crabtree's answer beares date Broughton Martij 18. 1640/41 in which hee delivers severall of his doubts concerneing Mr Gascoignes 5 opticall schemes & craves a solution, Shewes that it will not be possible to obteine the Moon's parallax by observeing the distance of any spot on her body from her limbe in the low & high altitudes, foreshews some oppotune times for observations of the planet Mars his place. & then proceedes —

> The Eclipse of the Moone 10 December 1638 I observed as exactly as the dim clouds would give me leave. the time I tooke by a good brasse clock whose minute finger goes about in a circle by it selfe once every hour & being fastned upon the axis of the lowest wheele is firme without any jetting. the circle is divided into 60 minutes & every minute into 3 partes & so is discernable by 10″ of time & lesse. I had one at the clock to note the time when I gave notice & by 10 severall observations of the lunar azimuth taken by a semicircle of 4 foot diameter I found the clock to keep pace with the heavens in a constant tenor & within lesse then a minute dureing all the time of the Eclipse. that difference which was is rectified in the following note [a table of observations of the 10 Dec 1638 [OS] lunar eclipse, from Broughton, near Manchester].

…

hee addes

> The observations differ a little from Phocilides, but agree with yours to admiration onely the correction of the clock either in you or mee is a little missed. for the observations allow no differences of Meridians betwixt us which by itinerary distance should seeme to be about 3 min, of time. For my parte I observed both the time & Eclipse with as much exactnesse & care as I could. The time I durst say much by. yet will not be too confident in any thinge.

I must not omit that in the very preface of this letter hee spends one whole page in lamenting the death of Mr Horrox whom thence I guesse to have died in February precedeing to wit $16^{40}/_{41}$ or 41.

10.
GASCOIGNE to WILLIAM CRABTREE
22 March 1641

Sources:

The National Archives, RGO 1/40, f.14v

I remember 2 or more mistakes in my last letter, one whereof was in the Sun's [semi]diameter Jan 20. upon reiteration of the calculation I find it 16′-24″. Also Feb

15 by the mentioned contraction of the diameter of the single convex glasse & an augmentation of the distance of the table to about 5 yards & to neare 7 yards I gathered semidiameters 16–20 & 16–22 & that which puts all out of doubt is I called to mind & examined the little table wherein I tooke the Suns diameter at the last Eclipse in Ma<–> & find it 149 such as the greatest is 154 - that is neare 32″ lesse than 16′-25″ the greatest: & besides this, this 10 of March I found it in a radius of 250 exactly answering to 17 seconds diminution I would gladly be certeine whether I in these 2 starrs have pitched on the right ones Lansberge & the observations differ so very much that I know not what to thinke.

The first that I sent was on Jan 11 when the Moon's lower limbe was 15′-19″ above the * & neare 2′ westward (I believe I wrote wrong before) this was very nigh the meridian therefore the time here 7^{h}-51′-56″. the moones center had passed the * 7′ – at the altitude of the great dog 14°-50′ = 8^{h}-03′ ap: time Lansbergs long. of the Moon differs 1°+ his lat 14′ lesse from the Bulls eye. –

Feb:9 at the great Dogs altitude 18°-10′ the Moon's Northwest limbe & the Star which I suppose the 13 of Orion were 15′-02″ distant at 36°-40′ altitude of Cor Leonis = * 8h-36′ the lunar Northwest limbe & * were in the same verticall the distance 16′-42″ (whence you may gather the inclination which I omitted in my last letter. This is practicable without the little inclinator. Lansberg & this differ in longitude some 40′+ in lat. some 5′...

11.
GASCOIGNE to WILLIAM CRABTREE
6 April 1641

Sources:

The National Archives, RGO 1/40, f.14v. No text of this letter was transcribed by Flamsteed.

Notes:

By J.Flamsteed: Mr Gascoines second letter beareing date Aprill 6: 1641 conteines nothing more then the explaineing severall schemes of his opticks in which Mr Crabtree found somethings not explaind to his capacity. Mr Crabtrees answer to both these as I conceive, is dated May 26. 1641.

12.
WILLIAM CRABTREE to GASCOIGNE
26 May 1641

Sources:

The National Archives, RGO 1/9, f.5v & 1/40, f.14v-f.15r.
Bodleian Library, MS.Eng.7031, f.32v-f.33r [source used]

Concerning the Ballance I am unresolved & unsatisfied, as also whether that can be true which you meane of any Cube of water to the same Cube of a heavier substance, I wish we could make some subtill experiment concerning it, I kindly thanke you for your frec [sic] courtesie, & the great paines you have taken with me & for me in explaining the grounds of your glasses & the manner of the falling of the

Sun's rayes, I am now pretty well satisfied about the 3 diag: of the reason why *egf* is equall to *icd*, And I thinke I rightly apprehend your meaning in most things, yet in some few I doe not fully apprehend your expressions. You say that the rays after they have passed through the convex glasse move like *cde* in the 2 & 4 diag: because they fall betweene *lt* of the 2 diag: whereas those of the 1st diag: fall upon *b* of the 2 diag: with an Angle greater then *lbt*, & therefore are forced towards *i* & soe are uselesse. I understand not this if your letters be placed as in that 2nd diag: you sent mee, unlesse you meane that the beams of the 1st diag: which come from any point within the Sun's circumference are refracted toward *i* & soe are uselesse as to the determination of the diamiter. For now I suppose I understand fully, how & in what manner the beames coming from the sides of the Sun, & falling upon every part of the convex are there refracted from every part of that Convex to the opposite side of the Image, and the beames coming from the Center of the Sun & falling upon every part of the Convex are thence refracted to the Center of the Sun's image. And those coming from any intermediate point in the Sun falling upon every part of the open convex are from each part thereof refracted to his correspondent place in the intermediate parts of the Sun's Image as I can now by drawing them at large satisfie my selfe fully, & some demonstrations in Scheiner compared with yours have both cleared my understanding & confirmed my judgement herein. I writ to you to know what you meant by the augmentation of the Solid by *ab* above *ql* as the Cube of sine 6° to that of 15′ e[c] [= etc?] (as you may see in my last letter) whether it hold not alsoe in lines, whereto you onely answer thus (I said solids because the increase in the Sun is as 1000 to 1) I beseech you explaine both unto mee. And here also give me leave to tell you how I conceave & understand you in some other things & if I be right let me know, if not I pray you pluck mee by the eare where I am amisse.

First I conceave that those convex glasses which have the greatest axis *WB* (diag: 1st) make the greatest distance of *iC*, & in that sence I suppose you meane it, when you say (in your last letter) that it is onely the forme that make the distance soe great in your great glasse.

Secondly I still perceive your opinion is that the glasses should not be segments of Circles but ovalls, that soe the point *m* (in diag 5) & *C* (in diag 1) might alwaies have one & the same distance from the glasse at every widenesse of the Convex, whereas when the Convexes are formed of segments of Circles it seemes those points are moveable.

Thirdly it seemes that in the same glasse, the least *iB* mak[e]s the greatest *iC*: for soe (in your former letter) you said that is [it?] was possible soe to diminish *iB* that *iC* (which was 160^{in}) should be 320^{in} or more & (in your last letter) you mention inlarging *fi* (in the Scheame on your margent there which is all one with *kC* in the 1st diag:) to 7 yards, or what length one will; Now if the proportion of *GF* to *DE* hold the same in every widenesse of the same glasse (which I conceave it must needs bee) then having the axis *WB*, & the widenesse *Bi* (Diag: 1) one may find the distance *iC* answering to any widenesse of *Bi* for

$$\left.\begin{matrix} WB \\ Bi \end{matrix}\right\} \begin{matrix} BWi \\ Wi \end{matrix} = WBA = iBg.$$

Then $GF : DE :: WBA\ (= BWi)\ DBc - iBg\ (= BWi) = icB$:

or $iBg + 90 - DBc = iBc : \left.\begin{matrix} icB \\ iB \end{matrix}\right\} ic$

Thus (if I understand these things aright) *iC* will be very litle changed though one alter *iB* a great deale, for if *iB* be 240, *WB* 2580 (as you assumed in your last letter & *DE*, *GF* also as you there tooke them) *iC* will be 4457 & if *WB* bee 2580 as before & *iB* but 5, *iC* will bee 4503, so that (if these be hundreth parts of an inch) *iC* will not alter above ½ an inch though *iB* be made 4 times wider in one observation then in the other. Soe if (in your great glasse) *WB* bee 8536, *Bi* 250, *iC* (if the glasse have the same proportion of refraction with the other) will be 14852. In the same glasse if *Bi* bee but 5, *iC* will be found but 14862, if I understand the rules & grounds aright, which insensibly differs from the former though the widenesse be 50 times lesse, I desire to have your opinion herein.

Notes:

By J.Flamsteed: In the preface of it I find this notable paragraph.

> The 6 of Aprill [JF: I suppose last past] the Sun being about 3 degrees high appeared tractable to the bare eye, through thin clouds & vapors; I fixed 2 small needles & went back exactly squarewise from the one of them till by the outside of the one & inside of the other (both wayes tried) I could just see the Suns transverse edges. I there noted the place of mine eye, which being after measured was *AB* 2011 such partes as *BD* (the distance of the needles from the out side of the one to the inside of the other) was 19.75 hence I gather the suns diameter 32–03 semid. 16-01½ The Suns limbe was not very exactly distinguishable because of its dimnesse in the cloudy vapors else I should have thought this an exact observation. for *AB* was about 5 yards distance:

- concerneing the appulse of January 11 he writes –

> 1640/41 Jan: 11 h7-52 (at Middleton) T.ap: h 7^h-57′. T.med.
> By Lansberg according to my calculations. Moon 4°.33′ Gemini Lat 1°-32′ A
> By Argolus in his Ephem: after his owne calcul. 4–50 Gemini Lat 1–20 A
> By Tycho according to Agolus & [Origanus] Ephe. 3–35 Gemini Lat 1–20 A

his next worke is to demonstrate the method of findeing the Suns distance by observeing the illuminated partes of the moone[36]. which is also done by Ricciolus & I therefore omit, hee writes thus concerneing the Solar Spot

> The motion of the Suns Spotts is nearest parallel to the Equinoctiall (you know) when they have theire greatest inclination to the Ecliptick toward the Equinoctiall which is the latter end of our November its an hard thing to designe the just day of the spot aequilibrium or theire greatest inclination, if it be at all possible, Scheiner designes the times of the spots aequilibrium about the 21 or 22 of August & February, Theire greatest inclination (about 7 deg) about the 22 May & Novembris Stilo Ang. – and yet as neare as I can guesse by compareing togeather some of his observations. The time of theire greatest inclination seemes rather to bee about the 28 of May & November. The aequilibrium about the 25 Feb & 28 August. St.Angl – The greatest inclination of the spots way to the ecliptick I cannot certeinly gather, observations make it variously But most betweene 7 & 8 degrees.

[36] the lunar dichotomy method of Aristarchus. This had been repeated by Gottfried Wendelin using a telescope and was reported in his *Loxias seu de obliquitate solis diatriba* (Antwerp, 1626), 10–11, Riccioli also mentioned this update of the method in his *Almagestum novum* (1651)

the rest of his letter onely conteines his further dissatisfactions & doubts conceroneing the opticks of Mr Gascoigne. with admonitions of severall opportunitys for celestiall observations which therefore are pretermitted.
Mr Horrox father died May 3. 1641: griefe for his sonnes hastning his owne death:

13.
GASCOIGNE to WILLIAM CRABTREE
22 June 1641

Sources:

The National Archives, RGO 1/9 & 1/40, f.15r-16r

[Note interpolated by Flamsteed: Mr Crabtree had lately written to him to observe the visible dichotomys of the Moon. hee returnes—]

I doe not approve of the conceipt, my reason is because I find the first illumination of divers Spots a long time to precede the true line of illumination as on the former mentioned 11 of Jan. within 4.6 of the Moon's upmost limbe was a parte like a ridge 0.85 in the darke illuminated. I take it to be better by the generall consent of observations taken as neare as may be from the line of generall illumination to fix upon the most frequent *Abc* (or angle subtended by the Moon's orbe at the Sun) I find that line much more unequall neare the 1st quadrature then a day after it. here are 2 other observations much more circumspectly made. Aprill 7 20.5 Moon Ray. 4500 (*qk* diag.5) 9h p.m. 95 parts illuminated. 1st 9.1½ alt Arcturus 35°.50 (i e) hor 8h 32′: April 9. 21.0¼ Moon R. 4494 (*qk*) 13.5¾ illuminated: at 9h 32′, 45°30′ alt of Arcturus. & 13.1 was illum: at 38°-42′ alt of Arcturus h.8.38. Bee pleased to let mee see what you gather hence

[Note interpolated by Flamsteed: after this hee proposeth a solar theory & (holdeing the obliquity of the Ecliptick mutable) an hypothesis to salve it. but these I find in one of his last letters hee retracted therefore I pretermit it. but may not this notable following paragraph.—]

There is not any thing more certeine then that the Sun's semidiameter changeth not above 35″ at the most. when you come hither wee shall review some certeine proofes of it, onely I dare not determine whether the greatest be 16′-25″ or 16′-30″ or the least 15′-50″ or 15′-55″ being both the least by the great glasse & the greatest by the least glasse. On Whitsun eve I observed Venus' diameter 1′-04″ by the brasses exquisitely.

I take it to be very little different from Mr Horrox & your observations reduced to the true diameter of the Sun, respect also had to Venus' place when I see it, it then wanteing some 16 dayes of the conjunction

14.
WILLIAM CRABTREE to GASCOIGNE
13 July 1641

Sources:

The National Archives, RGO 1/40, f.16r
Bodleian Library, MS.Eng.7031, f. 33r [source used]

I doe not yet apprehend how *iC* (diag:1st) may be found as truely by the concave as without it; nor what you meane by the diminution of *kq* or *BC* of the 1st diag: Nor

Fig. 18.9

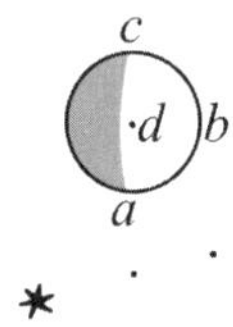

how (the concave of diag:2 being turned toward the Convex of diag: 1st) *ab* of the 2nd is neare equall to that diminution.

Neither can I hope with my irregular glasses to try any thing in the Sun to helpe me to any certainty herein; But if [I] perceive not your meaning perfectly at a further review of your lines, I shall hope to be fully satisfied by ocular demonstration when I see you, concerning the augmentation of the solid by *ab* above *ql*, I now receive both your meaning & the reason of it; And soe alsoe I perceive that though *iC* demonstratively & abstractively considered be litle altered by the diminution of *iB*, yet practically & sensibly a litle *iB* admits a great distance of *iC*, because of the acute intersection of the distance and diamiter determining beames. The precise point of which intersection (as you well demonstrate) is not sensibly determinable for a great distance.

15.
GASCOIGNE to WILLIAM CRABTREE
29 July 1641

Sources:

The National Archives, RGO 1/40, f.15v-16r

you may find the elevation of the pole in my note of the observation of the eclipse of the Sun 53°-41′. I find in my note of the last appulse, the star was Cor Leonis[37] how I wrote wrong I know not: the 1st observation 19¾ was betwixt the next illuminate horne of the Moon & the * like a * in the margin (*cab* being the illuminated parte of the Moon) [18.9].

The 4th observation *ca* * were all in the same line (not verticall) as neare as I could by *ca* direct a line to the *, the next day at 10h the clock was 37′ too late.

16.
GASCOIGNE to WILLIAM CRABTREE
1 Sept 1641

Sources:

The National Archives, RGO 1/40, f.16v [mention only]

Notes:

By J.Flamsteed: Mr Gascoignes letter of September the first conteined onely some conick problemes & no celestiall observation

[37] Flamsteed note: hee had wrote <Caudar>

17.
GASCOIGNE to WILLIAM CRABTREE
1 Nov 1641

Sources:

The National Archives, RGO 1/40, f.16v
[Not included in this selection]

Notes:

This letter includes a description and diagram of the telescope 'inclinator' that Gascoigne has devised, along with a few observations (diameters of the Sun, Moon, and Jupiter, and a position of Venus).

18.
WILLIAM CRABTREE to GASCOIGNE
1 Nov 1641

Sources:

The British Library, MS.Add.4021, Sloane Collection, ff.54r-62r
[only the last part of the letter is presented in this selection]

What Scaliger or others say about it I know not, however it is certaine that the Observations of the Equinoxes made by the Ancients were very grosse & uncertaine Ptolemy confesses to the ¼ of a day. if you compare those in Phocilides p45 you may see they ordinarily differ ¼ of a daye from that of the precedent yeare, & the first & last differ ¾ of a deg from the <tenor> that the[y] should hould, Kepler is of Opinion (Tab.Rud. p118) that Ptolemy erred a day in the account of his equinoxes in Numbringe the dayes of the <Egipt-> yeare which hee confirmes, from a place of <–>, Accountinge the Equinoxes a day latter then it were indeed, thus little Certaintye have wee.

Kepler promiseth a confirmation of his account of <–> & other thinges in a peculiar treatise but I know not whether it be extant. I should desire your Opinion conserning these thinges at some convenient leasure. It may be Lansberg in his Chronology may deduce some certaintie from Ellipsis. this Subject would require a large time to goe fully & Sufficiently through.

I have spent some dayes in the Examination of these thinges, but nott beinge Compleatly furnished with history I cannot satisfie my selfe in some thinges. I would desire to know what Chronologies you have.

If the Anntients bee thus uncertaint about the Equinoxes they cannot be much more certaine about the Obliquitie of the Ecliptick. Kepler tab.Rud: p116 <cals> it Res dubia sit, An omnino mutetur obliquitas Eclipticae successu saeculorum, and more fully p27 Ipsam scilicet Obliquitate Eclipticae olim fuisse majorem, res non tantum est dubia, sed validis argumentis ex ipsis veterum observationibus a me convulsa, & procul dubio plane falsa. yet in his Tables hee Expounds 4 severall hippotheses of the Obliquitie to anser Observations somewhat neare. In your Letter you put the greatest 24 30′ the least 23 30′ I could wish Observations would agrree with this after your Hipothesis. but certainl[y] by your Theory the Obliquitie alters slowliest when the <—> are about the Apogeum (if you be not satisfied of this wee shall dispute it the next time) And then wee must make the inclination of the Orbes almost 4 deg (as <–> judge I have not tried) to make it agree to observations. which would <–> the Ancient Equinoxes & Solstises.

(mentioned Theor. 7 & 8) to differ from the meane ones in the Creation about 6° <–> which is not probable: Wendelinus (as Master Foster saith) makes the greatest 24 30′ the least 23° 30′ hee makes it yet decreasinge till A° Chr. 1860, the <wh-> restitution <–40> Tropicall yeares. which as hee handles the Matter hee makes to agree with all Observations old and new for 2200 yeares within 1/6 of a minute: I am forced to conclude, else I loose this weeke also I shall exceedingly desire to heare from you, you may direct your Letters as before either to Richard Martincrofts in Manchester, or to the <Clerks> house in Rachdale those that come through Manchester are sometymes a good while before they come to mee.

I shall desire your opinion Conserninge the printinge of it, & where to put a period to it or what Epistle to prefix therto, I feare I have out tired your leasure with these many lines and therfore take leave restinge

yours to his little all

Wi: Crabtrie.

Notes:

The copy of this letter, though fairly lengthy, is fragmentary and suffers (as noted by Henry Power) from being 'so rudely…copyed out'. The letter is mainly devoted to a discussion of conic sections and the planetary motions, referencing the works of Mydorgius, Tycho, Hipparchus, Ptolemy, Phocilides, Kepler, Longomontanus, and Lansberge. Crabtree also describes Gascoigne's solar theory, as he understands it (a theory that Gascoigne later abandoned, according to Flamsteed). Ancient chronology is discussed and it seems that Crabtree considers it valid to take the assumed date of the Creation into account in his calculations. As an aside, he informs Gascoigne that 'In these Calculations I use Vlacc's Cannon of Sines & Tangents which are to everie 10″ & Wingates Logris for the Numbers'.

19.
GASCOIGNE / WILLIAM CRABTREE (?)[38]
Date unknown

Sources:

Bodleian Library, MS.Eng.7031, f. 41r

Most estimated friend,

In my last letter I promised in my next to answer to divers parts of yours in that unmentioned. First I can more easily increase then cleare your doubts about the anciently observed equinoxes. Scaliger as most others preferrs Hypparcus observations before Ptolomies & of those he saith,

Omnium accuratissima … est illa, qua equinoctium notatum est in tertia intercalarium: sequente quarta, in media nocte, anni 601 definentis a meridie Thoth primi Nabonassar, anno 32 tertiae periodi Calippicae, quod tempus congruit horis 12 a

[38] It is not stated in the original copy whether this is from Gascoigne to Crabtree or vice-versa. It is also undated, but has been placed here in the selection of extracts since it appears to be continuing the discussion about chronology that featured in Crabtree's letter of 1 November 1641.

meridie 26 Septembris anni 4567 periodi Iulianae, seria septima, sequente prima[39] 4567–764 (the root of the Julian Period) = 3803 (the Julian yeare of the world) - 601 (from Nabonassar) = 3202 . 127 dayes Phocid: p:80.[40] the difference of the Epoch of Nabon: & the world. But our Countryman Lidiat Scaligers firce Antagonist vilifies Hipparchus (yet he tells us of his vernall observations this not unworthy your reading – excessus supra 365 dies videretur tunc paulo major quinta diei parte sed et ex duabus primis observationibus factis circa vernum aequinoctium (quae quidem duae in vulgatis exemplaribus Graecis et Latinis perperam in unam confunduntur; at in peruetusto Latino optimae fidei manuscripto, quod mihi videre contigit recte distinguuntur uti constant ex collatione Ptolemaicae cum earum priori) constituitur annus dierum 365 et horarum duntaxat 5 nam cum anno precedente aequinoctium obseruatum esset mane sequente anno obseruatum est circa quintam hora diei.[41] And long after this he thus proceedes – Alfonsini, quos hactenus secuti sunt Pontificij, adimentes tantum 134 particulam diei mensurae anni Caesareae, dum Hipparchi et Albantenij obseruationes praecipue respexerunt, vernum aequinoctium Ptolemaicum anteuerterunt integro die aut amplius, omnino absq; ratione; quandoquidem Ptolemeus varijs instrumentis, quorum deterrimum unico illo Hipparchico esset longe melius aequinoctia indagauerit.[42] For the begining of the Period he writes thus: Initium vero periodi suae 36 annorum statuit Calippus quasi in eunte anno 3° olympiades 112 atque octauo anno 6i cycli Metonici - videtur tamen Calippus neutiquam fecisse periodi suae initium eo vel die vel mense uti Jos: Scaliger censuit, verum distulisse ipsum usq; ad quintum Nouilunium ejusdem anni quod contigit item tricesimo die mensis quarti septimanae quinto sub auroram.[43] He thus conjoynes his accounts 32 yeare of the 3rd Period of Calippus was Nabonazars 601yeare = 3858 of the world = 184 of the Grecian empire.

It is most certaine that the equinoxes are unsutable each to other & is probabline to the obliquity of the Zodiack is dubitable to at least 10′ from the uncertainty of their observed equinoxes.

20.
WILLIAM CRABTREE to GASCOIGNE
6 December 1641

Sources:

Philosophical transactions, v.30 (1717–19), 608–9. Extracts from Mr Gascoigne's and Mr Crabtrie's Letters, proving Mr Gascoigne to have been the Inventor of the Telescopick Sights of Mathematical Instruments, and not the French. By W.Derham, Prebend of Windsor, and R.Soc.Soc. [source used]

The National Archives, RGO 1/40, ff.16r-17r

---

[39] Scaliger, Joseph, *Opus de Emendatione Temporum* (1583), 309 (1629 edition)

[40] Holwarda, Johannes Phocylides, *Dissertatio Astronomica quae Occasione ultimi Lunarie anni 1638 Deliquii* (Franeker, 1640), 80

[41] Lydiat, Thomas, *Tractatus de varijs Annorum* (London, 1605), 59

[42] *Ibid.*, 331

[43] *Ibid.*, 171

That which you give me a full Projection of was above my Hope: and if the Screws keep an exact Equality of Motion forward in each Revolve, it is a most admirable Invention; and with the other Accomodations, I had almost said without Compare. But that the Divisions of a Circle should be measured to Seconds, without the Limb of an Instrument, or that Distances, Altitudes, Inclinations, and Azimuths should be taken all at one Moment, without the Limb of an Instrument likewise, and each to any required Number of Parts; or that the Diameter of Jupiter should be projected in such prodigious Measures as you speak of, &c. were enough to amuse and amaze all the Mathematicians in Europe, and may indeed be rather a Subject of Admiration than Belief, to any that hath not known your former Inventions to exceed Vulgar (I had almost said Humane) Abilities. And for my Part, I must confess Modesty so checks my ambitious Desires, that I dare scarce hope such Miracles should ever be produced in real Practice to such Exactness.

...no Man that hath written of the Diagram [of Hipparchus] understood it fully, or described it rightly, but only Kepler and our Horrox; for whose immature Death there is scarce a Day which I pass without some Pang of Sorrow.

Notes:

By J. Flamsteed: concerneing the observations of the Moon's enlightened partes hee writes

> your observation of the 10 Nov. 1640 makes the Sun distant 1519 semidiameters of the earth That of the 7 of Aprill 1641 h. 8^h 32′ make it distant 62600 that at 9^h. 1863 those of Aprill 9 extend it ad infinitum. making the 2 angles *acb* + *cab* above 180°. the first some few minutes the latter above 2 degrees. which great difference in that short space must needs arise from the erroneous observeing the Moon's enlightened partes & not from any error in the Moon's place.
>
> ...
>
> For the last observation October 30. 1641, at 7^h 56′ p.m. Altit. ocul. Taurus [the eye of the Bull] 18°-40′. At that time by Mr Horrox hypotheses the Moon 16° 51′ 41″ Aquarius Lat 5° 06′ 52″N long Moon visa 16-20-31 Aquarius latitude 4-20-17N Sun 17°-44′-37″ Scorpio visible distance Sun & Moon 88°-36′-10″.
>
> Whence hee deduces by processe the Suns distance 8844 semid.ter. & the horizontall parallax 22½″. hee addes
>
> But if the Moon's enligtened partes had beene but 13370 (& so but about 2″ difference from observation) the sunns distance would be 13550 semid.ter. parallax 15″.
>
> Or if the moones distance were but 8 min further from the Sun then Mr Horrox makes it viz 88°-44′-10″ all thinges else remaineing as observed the distance would be 13930 semid. Parallax 14½″.

21.
GASCOIGNE to WILLIAM CRABTREE
24 December 1641

Sources:

Philosophical transactions, v.30 (1717–19), 605. Extracts from Mr Gascoigne's and Mr Crabtrie's Letters, proving Mr Gascoigne to have been the Inventor of the Telescopick Sights of Mathematical Instruments, and not the French. By W.Derham, Prebend of Windsor, and R.Soc. Soc. [source used]

The National Archives, RGO 1/40, f.17r

Fig. 18.10

Mr Horrox his Theory of the Moon I shall be shortly furnished to try. For I am fitting my Sextant for all manner of Observations, by two perspicills with Threads. And also I am consulting my Workman about the making of Wheels like β, γ, δ, ε, of † Diagr.3, to use two Glasses like a Sector. If I once have my Tools in readiness to my Desire, I shall use them every Night. I have fitted my Sextant by the Help of the Cane, two Glasses in it, and a Thread, so as to be a pleasant Instrument, could Wood and a Country-Joiner or Workman please me.

Notes:

By J. Flamsteed: In a letter of Mr Gascoignes dated Christmas Grand Eve 1641 I find these observations thus delivered

> October 12. I observed 3 severall times Jupiter's diameter 503. 498. 503 &c

and after -

> It is true the Moon is allwayes very unequall in the illuminated line, to <–> quantity, it may be, scarce credible whereof here is an observation. No: 29 the Moon neare the Meridian & about $34^{4}/_{6}°$ alt. 13277 the Semid. 13993 the line of illumination beyond which apparently thus [see figure 18.10] were 3 illuminated partes affording 14842 for the greatest illumination whence wee may collect theire altitudes.

In a followeing paragraph relating to his former observation hee writes -

> I tooke the distance betwixt the glasse & pointers which I sent you, by compareing a few screw lines to the partes of an inch, by a payre of compasse & so erred. upon more diligent enquiry I find its logarithm 6.4622683 therefore the length 2859135.

Mr Crabtree [answered on] March 17 $164^{1}/_{2}$

22.
WILLIAM CRABTREE to GASCOIGNE
17 March 1642

Sources:

The National Archives, RGO 1/40, f.17r [a brief mention only]

Notes:

By J. Flamsteed: Mr Crabtree in his answer of March 17 164½ adverts that if that be the logarithme, the length is 2895913+ but in my tables that log. gives 2899135.

23.
GASCOIGNE to WILLIAM CRABTREE
date worn out
(after 17 March 1642 and before 21 June 1642)

Sources:

The National Archives, RGO 1/40, f.17r-

[Note interpolated by Flamsteed: concerneing the observations of the Sun's diameters Octob: 12 to satisfy Mr Crabtree hee writes thus in a following letter whose date is worne out]

The suns diameter October 12 was 27189 by the Screws & Wheels. at 3°-10′ altitude 25026 the contracted diameter at 11° alt 27023 contr. diam.

[Note interpolated by Flamsteed: & soone after hee adjoynes -]

so as the Suns Ray [radius] which I formerly gathered 15′-54″ was 16′-07$^{1}/_{3}$″ very little lesse then by the plaine or single glasse

[Note interpolated by Flamsteed: and soone after -]

The Sun's and Moon's diameters are of such a breadth that is a difficulty to marke both theire bounds at once, therefore an unit in Jupiter is as distinguishable as 10 in these. From this yeare & the last Hortensius is most manifestly in an error, the Sun in all the varietyes I have yet observed never fully reaching ½ of Lansbergs excentricity First <his> owne way by the usuall glasse and an obscure roome it is in the line brea[d]th

as 1127:1087::985″:950″ the diff: 35″ by the great case & one glasse

as 346:334::985:951 the diff 34″ by the old case & the glasse you see

as 25¼:26.2::949:985 diff 36″ both by the eye & by transmission

what greater proofes can be made I know not I esteeme the suns greatest diameter 16′ 25″, the least 15′ 51″ as heretofor observed.

[Note interpolated by Flamsteed: And a little after -]

if the times of these observations of the Moon's diameters by one great glasse agree as well with Mr Horrox hypotheses as their alterations doe it is more then likely to be true.

I found the Moon's semidiameter at the full in 1641 June 12 17′.9 & the least full Moon neare 15′.48 - ..1 17.9:15.48::63:54.5 + 8.5 = 63 alteration in the full Moon about 8.5 semidiameters of the earth such as the farthest is 63.

In the <-> October 11 1640 the Moon ½ diameter 16′-36″ = 0.6 & September 17 (1641) 15′.17 Moon Ray.

17.9:16.6::63:58.4 }

17.9:15.17::63:53.4 }5.0 precise difference of 5 semidiameters of the earth in Quadratures. different altitudes may cause some difference which I cannot certeinely correct & therefore for the present thinke this sufficient; onely it seemes the change in the full Moon is above 8. by my great glasse. & the noted altitudes I shall here after gather greater certeinety. It seemes Mr Horrox meane ½ diameters of the Sun & Moon are neare 53″ too little. for as I find the Sun's meane semidiameter 16′-08″-53″ = 15′-15″. Mr Horroxes Moon meane ½ diam. 16′-17″ - 47″ = 15′-30″ Mr Horroxes meane ½ diam Moon. so as the meane diameters are either as hee assumes them exactly equall or a very few seconds different.

[Note interpolated by Flamsteed: Mr Crabtrees returne was dated June 21 1642, in which I find an account of the lunar Eclipse Aprill 4 precedeing & a description of Mr Horrox theory of the Moon & correction of the Rudolphine numbers for calculations]

24.
GASCOIGNE to WILLIAM CRABTREE
date worn out

Sources:

Philosophical transactions, v.30 (1717–19), 605. Extracts from Mr Gascoigne's and Mr Crabtrie's Letters, proving Mr Gascoigne to have been the Inventor of the Telescopick Sights of Mathematical Instruments, and not the French. By W.Derham, Prebend of Windsor, and R.Soc. Soc. [source used]

I have given order for an Iron Quadrant of Five Foot, which will give me the 1000th Part of One Degree, which shall be furnished like my first Scale; only my Workman is so throng for my Father, that I fear it will not be finished before the Eclipse. I have caused a very strong Ruler to be exactly made, and intend to fit it with Cursors of Iron, with Glasses in them and a Thread, for my Sextant.

Notes:

By W.Derham: labelled '10th letter' in Crabtree's hand.

25.
GASCOIGNE to WILLIAM CRABTREE
8 June 1642

Sources:

The National Archives, RGO 1/40, f.20v

[Note interpolated by Flamsteed: In one of Mr Gascoignes letters dated June 8th I find his method used to gaine the picture of the eye. Which he thus describes]

I after no few plotteings find how to performe this have at last thought upon & practised this simple yet prime way when the season serves for it. I take a fresh eye in a frosty evening & place it where it may be frozen thorow. then I can cut it as I please and each part will be very different in colour, & remaine in theire naturall site which may be pricked forth by an oiled paper. or the halfe eye being fastened to a rimmeing table so as the cut side be from the table (moveing on<->[])] whereon another rimmer beareing a glasse convex of the one or both sides with a thrid haveing a table covered with white paper are likewise moveable, then alloweing the Sun to shine through an hole of 3, 4 or 5 inch breadth into a roome otherwise darkened turneing the cut side of the ey to the Sun passeing by the other 2, you may by removeing the table covered with paper find the true place of representation, which will if your glasse be great & good appeare not discerneably infferior in appearance to its object, & for magnitude thus. If the distance betweene the ½ eye and glasse exceed that of the glasse & table the representation will be lesse for extent but brighter for colour. If otherwise the contrary. When you have this as great as you please you may by your pen draw each part upon the paper. Or at any time I can describe the prime parts thus having a glasse & table fitted to observe the Sun's spots I place an eye with the horny tunicle either upward or downward betweene the inmost glasse & the table. So neare the glasse as your eye will allmost fill up the compasse of the Sun's image. Then the representation of the eye will be large (proportionable to the Sun's image) upon the table and thus I pricked out these 3 figures of the horny tunicle Christalline outward & inward humor, which is for ought I yet know wholly unknowne to the world (and which I first observed by a frozen eye, after upon triall in all dried Christallines so that theire are fower differences or humors) the outward Christalline I place as even as I can in a perforated thin plate whereon I spread the ciliary processe which keepe the Christalline right <et>

[Note interpolated by Flamsteed: the scheme of the eye is wanting therefore I omit those measures of it which hee tooke, by observeing it in his long tube, with his screw. Hee subjoynes as the summe of all.]

By these pointes ADBLEC and the cited prop of Midorg you may find that the horny membrane of the eye is more compound then the Circle Parab[ola]. Hyperb[ola] [or] Ellip[se]:

Because Scheiner accounts it an impossibility to find out the refraction of any of the 3 humors I have bestowed some searches & find by undeniable proofe that the glassy humor which they proportion like glasse hath the same with water. I can as easily prove that of the watry humor. the other in the comrehendeing & comprehended christallines are more difficult because of the 4 severall refracteing superficies, and theire unknowne formes so that I almost despaire of finding them. Hence it followes that every ray of incidence is seven times refracted in the eye before it reach the bottom.

[Note interpolated by Flamsteed: I may now enquire who first found this best way of dissecteing the eye: & I could wish the schemes hee mentions had not beene wanteing so had wee certeinly enjoyd a far more accurate description of this naturall telespore then any author yet hath given us or wee may expect, except from the like experiments. this as haveing no relation to his Astronomy I have here transcribed by it selfe for I thought it not meete to leave out a passage which for ought I yet know might vindicate the invention to my author. or at least shew the sublimity of his wit & profoundnesse of judgement.]

26.
WILLIAM CRABTREE to GASCOIGNE
21 June 1642

Sources:

The National Archives, RGO 1/40, ff.17v-19r
Opera Posthuma of Jeremiah Horrocks (ed. J.Wallis) (London, 1672–3), 467–472
[omitted from this selection]

Notes:

This lengthy letter is primarily devoted to explaining Jeremiah Horrocks's theory of the moon and how to apply appropriate corrections to Kepler's Rudolphine Tables. It also includes some observations of the lunar eclipse of 4th April preceeding. These eclipse observations are included (in Latin) in Flamsteed's *Historiae Coelestis* (v.1, p.4), but the date of the source letter is there given as 22 June 1642. In Horrocks's *Opera Posthuma* the date of the same letter (also in Latin translation) is given as 21 July 1642.

27.
WILLIAM CRABTREE to GASCOIGNE
"28 December 1644"

Sources:

Bodleian Library, MS.Eng.7031, f.29r

First in the 1st diagram I should desire you explaine in what case or example the Ray *AB* may be said to fall upon the point *B* & to be a parallell ray of incidence & how you find *kC* or *iC* & *qi* which it seems by the demonstration you take as given; I conceave *AB* a parrellell ray of incidence refracted in *B* & making its concourse in *C*, so *zm* a ray of the object *z* refracted in *m* & *b* & concuring in the point *q*: but yet I know not how to find *iC* or *iq*. In the 2nd diagram I conceive *AB* to be againe a parrellell ray which refracted in the point *B* of the concave glase paintes the semidiameter *gD* at the distance *kg*, & soe *a*, to be either an object in the free Aire neere the glasse: or rather the Sun shining through the convex glasse at *a*, whose beame *ab* refracted in the pointes *b* & *c* paintes it semidiameter *dh* (at the distance *kh*) & *eg* (at the distance *kg*).

But I conceive not the reason how the Sun beames should passe unrefracted through the convex glasse (supposed as I conceive, to be) at *a* in the 2nd Diagram, & soe should make the true Angle of the Sun's semidiameter *bal*: neither doe I understand why in the 3rd Diagram the ray *gh* (if it be a ray coming from the limbe of the Sun) should be refracted from *g* just to point *c*. Nor how the Angle *dci* not withstanding the refraction comes to be exactly equall to *gef*: I pray you therefore helpe me out here for I can make nothing of it: I am at the like stand alsoe in your 4 & 5 Diagrams in both which the Sun's rayes seeme to come through the convex glasse in the *k* as from a point unrefracted, I pray you therefore give me some demonstration or direction how this comes to passe & apply it to the 5 Diagram where both ye glasses are convex. I pray you also tell me whether that 5 Diagram doe not represent your lesser glasse which you shewed me, & whether the convex glasse, *ab*, in the 5 Diagram nearer the eye be not composed of 2 convex segments the flat sides clapped together, & whether it be convenient they should be segments of equall sphires: & (if I might obtaine soe much of you) where & how you place a haire in your glasse, & for what uses: Soe shall I always acknowledge you my sole instructor in these optick secrets, & if it ly in my power be ready to requite soe great courtesies: & lastly I thanke you kindly for the draught of the scale, it is a most exquisite device if it be wrought & devided exactly.

Notes:

This letter is described in the Bodleian manuscript as 'Mr Crabtrees desires in his letter Deceb 28 1644'. This is clearly an error: Gascoigne was dead by that date. The content, moreover, is more akin to the subject matter of much earlier letters. It seems likely that this extract is from the concluding part of Crabtree's letter of 28 December 1640 (see letter 6 above).

Index

D. Sellers, *In Search of William Gascoigne: Seventeenth Century Astronomer*, Astrophysics and Space Science Library 390, DOI 10.1007/978-1-4614-4097-0,

Printed by Printforce, the Netherlands